Green Research, Developments and Programs

Green Research, Developments and Programs

Green Technology for Bioremediation of Environmental Pollution
Bana Bihari Jana, Dr. Jatindra Nath Bhakta,
and Susmita Lahiri (Editors)
2019. ISBN: 978-1-53614-528-1 (Hardcover)
2019. ISBN: 978-1-53614-529-8 (eBook)

Greenhouse Gas Emissions and Nitrogen Losses from Grazed Dairy and Animal Housing Systems
Jiafa Luo (Editor)
2017. ISBN: 978-1-53611-100-2 (Hardcover)
2017. ISBN: 978-1-53611-258-0 (eBook)

EPA's Clean Power Plan: Highlights and Implications
Joshua T. Graham (Editor)
2017. ISBN: 978-1-63484-862-6 (Softcover)
2017. ISBN: 978-1-63484-863-3 (eBook)

Deciphering Organic Foods: A Comprehensive Guide to Organic Food Production, Consumption, and Promotion
Darrel D. Muehling and Ioannis Kareklas (Editors)
2017. ISBN: 978-1-53610-517-9 (Hardcover)
2017. ISBN: 978-1-53610-524-7 (eBook)

Ionic liquids for Green Energy Applications
Elaheh Kowsari (Editor)
2016. ISBN: 978-1-63484-359-1 (Hardcover)
2016. ISBN: 978-1-63484-392-8 (eBook)

More information about this series can be found at
https://novapublishers.com/product-category/series/microbiology-research-advances/

Teena Mishra, PhD

Designing Efficient Utilization of Energy Systems

From Green Perspective

Copyright © 2022 by Nova Science Publishers, Inc.

All rights reserved. No part of this book may be reproduced, stored in a retrieval system or transmitted in any form or by any means: electronic, electrostatic, magnetic, tape, mechanical photocopying, recording or otherwise without the written permission of the Publisher.

We have partnered with Copyright Clearance Center to make it easy for you to obtain permissions to reuse content from this publication. Simply navigate to this publication's page on Nova's website and locate the "Get Permission" button below the title description. This button is linked directly to the title's permission page on copyright.com. Alternatively, you can visit copyright.com and search by title, ISBN, or ISSN.

For further questions about using the service on copyright.com, please contact:
Copyright Clearance Center
Phone: +1-(978) 750-8400   Fax: +1-(978) 750-4470   E-mail: info@copyright.com

NOTICE TO THE READER

The Publisher has taken reasonable care in the preparation of this book, but makes no expressed or implied warranty of any kind and assumes no responsibility for any errors or omissions. No liability is assumed for incidental or consequential damages in connection with or arising out of information contained in this book. The Publisher shall not be liable for any special, consequential, or exemplary damages resulting, in whole or in part, from the readers' use of, or reliance upon, this material. Any parts of this book based on government reports are so indicated and copyright is claimed for those parts to the extent applicable to compilations of such works.

Independent verification should be sought for any data, advice or recommendations contained in this book. In addition, no responsibility is assumed by the Publisher for any injury and/or damage to persons or property arising from any methods, products, instructions, ideas or otherwise contained in this publication.

This publication is designed to provide accurate and authoritative information with regard to the subject matter covered herein. It is sold with the clear understanding that the Publisher is not engaged in rendering legal or any other professional services. If legal or any other expert assistance is required, the services of a competent person should be sought. FROM A DECLARATION OF PARTICIPANTS JOINTLY ADOPTED BY A COMMITTEE OF THE AMERICAN BAR ASSOCIATION AND A COMMITTEE OF PUBLISHERS.

Additional color graphics may be available in the e-book version of this book.

Library of Congress Cataloging-in-Publication Data

ISBN: 979-8-88697-051-7

Published by Nova Science Publishers, Inc. † New York

Contents

Preface vi
Acknowledgments ix
Abbreviations xi
Chapter 1   Introduction to Energy Utilisation Systems 1
Chapter 2   Integrated Energy Systems 17
Chapter 3   Green Energy 37
Chapter 4   Sources of Green Energy 53
Chapter 5   Design of Sustainable Energy Systems 65
Chapter 6   Methods and Tools of System Design 77
Chapter 7   Building a Sustainable Energy System 95
Chapter 8   Efficient Utilisation of Energy Systems 107
Chapter 9   Green Technology 123
Chapter 10   Future of Energy Systems 147
References 159
About the Author 165
Index 167

Preface

Energy is an essential component of all life. The increasing demand for energy for human life and its impact on the environment necessitate a rethinking of the energy use system. Green energy is the solution to many of the environmental issues that we face today. The book concentrates on the notion of green energy and its application.

The book is divided into ten chapters. Chapter 1 provides an introduction to the energy utilization system, which includes the energy utilization system, green utilization system, urbanization and sustainability, and current challenges of energy and energy utilization. Chapter 2 confers integrated energy systems, which comprise energy system integration, development of an integrated energy system, reasons for integrated energy system development, sector integration, and challenges of an integrated energy system.

Chapter 3 provides an overview of green energy. It describes the difference between green energy, clean energy, and renewable energy; types of green energy; green energy and sustainable development; implementing strategic strategies; obstacles for the development of renewable energy; and the future of green energy. Chapter 4 highlights sources of green energy, which are solar power, wind power, hydro power, geothermal energy, biomass, biofuels, and tidal energy. It also elucidates the advantages and applications of green energy. Chapter 5 describes the design of a sustainable energy system and discusses global action for sustainable energy. It also highlights green building programs and sustainable design in buildings.

Chapter 6 provides an overview of the methods and tools of system design. This chapter presents energy system planning; renewable energy design; application of tools in decentralized energy planning; hybrid energy tools; agent-based modelling and decision support tools. Chapter 7 stresses building a sustainable energy system, sustainable design and building energy systems, green building, sustainable energy agenda, business sustainability concepts.

Chapter 8 describes the efficient utilization of energy systems, energy generation and distribution, ways to attain greater energy efficiency, energy efficiency and conservation, modern appliances, and building designs.

Chapter 9 describes green technology, its basic principles, and examples, such as LED lighting, solar panels, wind energy, composting, electric vehicles, wastewater treatment, and so on. It also explores the use of green technology, green machinery, and alternative technology sources. Finally, Chapter 10 discusses the energy system's future, including stages toward a future energy system, strategic implications, sustainable energy transition, and future problems, such as digitalization and the mission of decarbonization.

Dr. Teena Mishra
PhD in Management

Acknowledgments

I would like to thank Mr. Thomas Mazzaferro, as well as the editors and other support personnel at Nova Science Publishers, for their assistance during the book's development process. I would like to thank God for providing the opportunity to publish my work in such a high-quality publication. I would also like to thank my mother, who has been a source of inspiration in my life. Finally, I'd like to express my gratitude to my father, who assisted during the development of the book.

Abbreviations

AASS	Association for Awareness of Social Systems
ABM	Agent Based Modelling
AHP	Affordable Housing Project
AHP	Analytical Hierarchy Process
B2B	Business to Business
BECCS	Bioenergy with Carbon Capture and Storage
BIM	Building Information Modelling
BREEAM	Building Environmental Assessment Method
BTU	British Thermal Unit
CEA	Central Electricity Authority
CHEC	combustion and Harmful Emission Control
CHP	Combined heat and power
COE	Cost of Energy
COP21	Paris Climate Conference
DERs	Distributed Energy Resources
EISA	Energy Independence and Security Act
ELECTRE	Elimination and Choice Translating Reality
EPAct	Energy Policy Act
ESI	Energy Systems Integration
GBC	Green Building Council
GBI	Green Building Index
GDP	Gross Domestic Product
GSAS	The Global Sustainability Assessment System
H2S	Hydrogen Sulfide
HVAC	Heating, Ventilation, Air Conditioning
ICEs I	nternal Combustion Engines
IEA	International Energy Agency
IES	Integrated Energy Systems
ISO	International Organization for Standardization
IT	Information Technology

LACE	Levelized Avoided Cost of Electricity
LCA	Life Cycle Analysis
LCC	Life cycle Costing
LCOE	Levelized Cost of Electricity
LED	Light-emitting diode
LEED®	Leadership in Energy and Environmental Design
MCDA	Multi-criteria Decision Analysis
MNRE	Ministry of New and Renewable Energy
MPEB	Madhya Pradesh Electricty Board
MW	Megawatts
OECD	Organisation for Economic Co-operation and Development
P2G	power-to-gas
P2H	power-to-heat
PPE	Personal Protective Equipment
PROMETHEE	Preference Ranking Organization Method for Enrichment Evaluation
REN21	Renewable Energy Policy Network for the 21st Century
RESOP	Renewable Energy Standard Offer Program
STS	Science and Technology Studies
UN	United Nation
UNEP	United Nations Environmental Program
UNFCCC	United Nations Framework Convention on Climate Change
WCG	Western Cape Government

Chapter 1

Introduction to Energy Utilisation Systems

Abstract

There are various forms of energy, but the thing is to identify and utilise them properly. The increasing demand for energy and its influence on the environment inspires the need to rethink the energy utilisation system. Therefore, the energy utilisation system should be such that utilisation is more and wastage is less, reducing adverse environmental impact. The green energy system is a new concept reinvented to prevent the environment from being damaged and applicable in various sectors like electricity, transportation, industries, domestic, business organisations, etc. This chapter introduces the concept of the energy utilisation system and the significance of the green energy system. Furthermore, this chapter emphasises the introduction to the energy utilisation system and how to boost the use of green energy.

Keywords: green energy, renewable energy, energy utilisation system, fossil fuel

1. Introduction

Green energy is the theme that is gaining attention in today's world. Due to the need for modern energy sources, this subject is the centre of attraction for many investigators and practitioners. Energy is a fundamental element in human life. There is no agricultural, industrial, domestic, health, or other types of process that do not need a degree of external energy. Humans consume around 2,500 kilocalories of energy per day as food (Current Challenges In Energy, Openmind" 2022). Therefore, it is significant to produce and consume energy properly. The meaning of the right way here is the way that makes the least or no adverse impact on the environment. According to the Merriam-Webster dictionary, the word environment means the situations, conditions, or objects by which one is surrounded ("Definition of ENERGY" 2019). It contains animals, plants, humans, water, soil, air, and forests.

Burning fossil fuels (Figure 1) send greenhouse gases into the environment and contributes to global warming. Generally, climate scientists agree that the earth's average temperature has increased in the past century. If this tendency continues, sea levels will rise, and scientist predicts that floods, heat waves, droughts, and other extreme weather situations could occur more often. In addition, when fossil fuel is burnt, other pollutants are released into the soil, air, and water when fossil fuels. These pollutants contribute to air pollution: acid rain from nitrogen oxides and sulfur dioxide damage plants and fish. In addition, nitrogen oxides result in smog (Laxmi, Dudi, and Shika, 2015).

Figure 1. Burning fossil Fuels.
Source: https://cdn.pixabay.com/photo/2010/12/16/14/44/fire-3593_960_720.jpg.

Green energy is the solution to various environmental problems we face today. In the world, most of the energy is generated from non-renewable and can be re-produced over a short period, for example, oil, natural gas, nuclear, and coal. The advantage of this energy is that it is ready to use and cheap. The disadvantage is that they are finite and will sometimes end in the future. Moreover, this energy is also responsible for global warming, human health, and climate change (Laxmi, Dudi, and Shika, 2015). Clean energy sources are used extensively and come into usage more efficiently. It is not easy to use natural resources for energy implementation. However, it is the foundation of future energy needs. It is a fact that the natural essence of energy utilisation is decreasing adverse influence on nature and human society. Thus, characteristics in the deliberation of practical designs and operations are necessary (Current Challenges In Energy, Openmind" 2022).

In manufacturing and industrialised countries, the average daily quantity of extra energy consumed in collective activities, for example, domestic, industrial, transportation, etc., is equivalent to 125 000 kilocalories per person, which is fifty times more. In the United States, the number is one hundred times more. There is increasing industrialisation, and rapid infrastructure transformation tends to rethink environmental sustainability. Unplanned development of infrastructure, energy production, and utilisation deteriorate the environment. Energy usage tends to be more natural to develop a pleasant-sounding world where human development is well-suited to environmental sustainability (Zhang 2017). Natural sources of energy are the best means to protect the environment.

The global population is increasing and hence energy consumption. The ability to provide and supply advanced clean energy is becoming a significant technological and political challenge of the 21st century. Building and construction sector is one of the critical uses of energy. United Nations Environmental Program (UNEP) and International Energy Agency (IEA) report building and construction and operation sector accounted for 36% of global final energy usage and 40% of energy connected to CO_2 emission 2017 (UNEP 2018). The quantity and building energy demand vary by location, weather, climate, local, national, and occupant behaviour. Gross domestic product, gross national income, or purchasing power parity per capita are the methods to measure a country's wealth and impact energy usage in general, explicitly building energy (Stagner and Ting 2021). According to Ayhan Demirbas (2020), energy impacts all aspects of modern life. There is an increasing demand for energy due to the growth of the world population. Advanced and modern energy efficiency technologies reduce the energy needed to provide energy services and the cost of using energy and increase its reliability. Researchers focus on identifying new and renewable energy sources (Laxmi, Dudi, and Shika, 2015). Thus, there is increasing research on finding methods and techniques to generate and utilise energy, creating a greeenization. Greenization has become the top priority in various stages of energy utilisation, comprising the system, source, and service (Zhang 2017).

2. Energy Utilisation System

The term energy utilisation is an integration of two words one is energy, and the other is utilisation. According to Cambridge Dictionary, energy is 'the ability and power to be physically and mentally active' ("Energy" 2022). It means anything that can control people or any event to be active. Generally, energy word implies the ability to remain active. The word energy is derived from the Greek word energeia, which means operation or activity. Therefore, energy means active or effective. En means in, and ergon means work. Specifically, energy work includes the power that comes from various sources such as electricity or heat or light to do work. Erg is the centimetre-gram-second unit used to measure the quantity of energy or the amount of work performed. ("The World's Fastest Dictionary | Vocabulary.Com" 2022).

According to Collins dictionary, 'Energy is the capability and strength to do active physical things and feeling that you are full of physical power and life' (Energy Definition and Meaning | Collins English Dictionary," n.d. 2022). Various definitions and explanations define energy in its way. Simply, energy is the capability to perform the work, and the ultimate result of energy is the measurement of work done. Energy is the form that is converted into various forms and can be transferred to other forms or objects. Thus, it is not formed nor destroyed. Energy not only drives the development of human civilisation by its power but also worsens the world's sustainability through its environmental problems. Thus, active efforts should be made to develop a pleasant-sounding world where human development is well-matched with environmental sustainability (Zhang 2017).

India's declaration that it aims to reach zero emissions by 2070 and to meet 50% of its electricity need from natural energy sources by 2030 is an enormously significant moment for the global fight against climate change (Laxmi, Dudi, and Shika, 2015). India has created a revolution to develop new economic development that could evade the carbon-intensive approaches that various developing countries have followed. In India, there is a speedy transformation in which construction of new buildings, transportation, network, and factories. Oil and coal are the foundation of India's industrial growth and modernisation, allowing modern energy services (Laxmi, Dudi, and Shika, 2015). Fossil fuels are a non-traditional source of energy and are limited in quantity. Thus, it is necessary to use renewable energy resources. Some kinds of renewable resources, including wind power, hydropower, solar photovoltaic, solar thermal, air source, and geothermal, should be extensively exploited for electricity generation to substitute the usage of fossil fuels

("Current Challenges in Energy | Openmind" 2022). The natural forthcoming energy utilisation is a solution to provide greater sustainability in our power grid. The united state energy market offers a variety of products and services with green energy, also referred to as green power. Green energy is interchangeable with renewable energy, but there is a difference between them ("What Is Green Energy? Renewable Energy Source" 2022).

3. Green Utilisation System

There are two sources of energy; renewable energy and non-renewable source of energy. Renewable sources are energy produced directly from nature, for example, sun, wind, rain, and tides. It is probable to make it over and over when it is needed. It is also known as clean energy, and it is in abundant form. Renewable sources are biomass, tidal energy, geothermal, solar, tidal, and hydro energy. Non-renewable sources are not environmentally friendly and adversely impact the climate and health (Laxmi, Dudi, and Shika, 2015). Based on the efficient renewable energy bases are ranked as follows: ("What Is Green Energy? (Definition, Types and Examples)" 2022).

1. Wind power
2. Hydropower
3. Nuclear
4. Geothermal
5. Solar power

The world is moving from conventional sources of energy to non-conventional sources of energy to safeguard the environmental system. In the world, most of the energy is generated from non-renewable and can be re-produced within a short period, for example, natural gas, coal, and oil. The advantage of this energy is that it is a cheap source of energy, and it is ready to use (Laxmi, Dudi, and Shika, 2015). In the past mobilisable energy sources were animal, wind, human, biomass, and energy hydropower. Renewable energy has been a primary source in history, and wood was used for water, cooking, space heating, and water. Hydropower and conventional biomass are significant factors in the world's energy mix, contributing around 18% of total energy needs, whereas renewable contributes about 2% of the current world's primary energy usage (Laxmi, Dudi, and Shika, 2015).

Figure 2. Renewable Energy.
Source: https://cdn.pixabay.com/photo/2021/05/22/01/29/renewable-energy-6272343_960_720.png

According to the Environmental protection Agency, green energy offers the highest environmental advantage and comprises wind, solar, geothermal, biogas, biomass, and low-impact hydroelectric power. These renewable energy sources originate in the power grid and are indifferent to traditional energy sources when anyone flips a light switch or charges a phone. However, renewable energy (Figure 2) incorporates the same sources as green energy. This energy primarily includes technologies and products that can substantially impact the local and global environment. When anyone purchases green power, they also support various green energy projects and invest in technologies that assist them in growing ("What Is Green Energy? Renewable Energy Source" 2022). The assessment of the United Nations Framework Convention on Climate Change (UNFCCC) is that due to the insufficiency of the mitigation efforts, the temperature is probable to 3-degree Celcius by 2050, which is a serious matter to discuss. It would be a loss for agriculture and farmers. Let us talk about China which has emerged as the world's biggest maker of wind turbines and a producer of solar panels. It is building efficient kinds of coal power plants. Today China has 9GW of nuclear power, and in 2008, renewable industries created 1.12 million new job opportunities. Denmark shows the method for developing a low-carbon economy through the innovative application of green technologies. It gets the bulk of its electricity from coal. Due to efforts, Greenhouse gas emissions reduced by 14% during the last 20 years. It is the most influential European country, with just a 5 million population, and has some leading biofuel, wind cooling, heating, and efficiency worldwide (Laxmi, Dudi, and Shika, 2015).

There are several questions in front of researchers and scientists working on green energy. The first question is, can green energy be able to replace fossil fuels? The second question is whether green energy is economically feasible, and the third is which type of green energy is more efficient. The fourth question is: how does green energy help the environment? Researchers are focusing on finding relevant answers to these questions and focusing on green energy (TWI, n.d.).

Let us discuss all these questions:

- Can Green Energy Substitute Fossil Fuels?

Green energy can substitute fossil fuels in the future, and it may require varied production from various sources to achieve this. By bringing many green energy sources to meet the needs and advancements that are made concerning the development and production of these resources, there is every reason to believe that fossil fuels could be phased out. It is necessary to improve the environment, change and reduce climate change, and move further sustainable future (TWI, n.d.).

- Can Green energy be economically feasible?

The economic feasibility of green energy needs a comparison with fossil fuels. Fossil fuels are more expensive, but the cost of greener energy sources is cheap and will decrease over time. Other factors also favour green energy, such as the capability to produce relatively inexpensive localised energy solutions such as solar farms. In addition, green energy solutions' investment, interest, and development reduce costs. Therefore, green energy is not only economically feasible but also an ideal option.

- Which Type of Green Energy is Most Efficient?

Efficiency in green energy is reliant on location. If the location has frequented and intense sunlight, it is easy to develop an efficient energy solution. It is essential to analyse the entire life cycle of energy sources to compare various energy types, which means assessing the energy to produce green energy resources. The energy that prevents the environment to damage is purely green energy (TWI, n.d.).

- How green energy assists the environment?

Green energy is helpful for the environment as power originates from natural resources such as wind, water, and sunlight. Producing energy with a zero-carbon footprint is more eco-friendly for the future. It is used to meet the power needs for industries and transportation and can reduce environmental influence (TWI, n.d.).

Above all, questions are relevant and significant as the world faces energy problems. It is essential to utilise energy and reduce the adverse impact on the environment, and human energy is required to think in this direction to achieve sustainability.

4. Urbanisation and Sustainability

In the 20th century, urbanisation increased, which means the movement of the population from rural cities to urban cities. 56% of the global population is in urban cities compared to 29% seven decades ago, and the percentage is likely to increase to 68% by 2050, according to the United Nations 2019 report (our world in data cited in "Energy UtilisationPromotion of Energy Saving Policy n.d.). Energy utilisation in developed countries is more well-organised and efficient than in developing countries. Energy consumption among Organisation for Economic Co-operation and Development (OECD) nations is 25% per unit of GDP of non-OECD nations. In addition to energy challenges for building and construction, other components, such as urbanisation and sustainability, need to be addressed. Due to the increasing population and commercial level, the relation between energy intensity and urbanisation has become a widespread academic focus. Therefore, countries with a lower urbanisation level need to implement energy-saving policies. They prohibit rising energy pollution while increasing the level of urbanisation. It is important to use clean energy and other pollution-free resources with the increasing urbanisation to decrease energy pollution (Zhu et al., 2021). Developed countries provided guidance and policies to select energy-efficient products such as top runner systems, energy management systems, and labelling standards. The energy system of Japan indulges large consumers, including offices and factories, to set up energy-saving actions and measures. Developed countries should improve energy-saving policies, an international standard for energy-efficient products, and energy consumption control systems. In contrast, developing countries need to fight fundamental challenges such as regulations and laws which improve the awareness of energy saving among customers and laws ("Energy UtilisationPromotion of Energy Saving (Policy)," n.d.).

Innovations in energy enhance the process of urbanisation and also decrease the energy pollution produced by cities. Thus, to protect the ecological environment and prevent excessive energy consumption, which hinders the development of urbanisation, the state vigorously boosts innovation. Heavy energy consumption in urbanisation development requires innovation to conserve the environment and stop excessive energy consumption. In this manner, the countries not only improve the national level of innovation but also develop urbanisation while protecting the environment (Zhu et al., 2021).

From 1980 to 2010, Salim Ruhul, Rafiq, and Shuddhasattwa Rafiq investigated the effects of renewable, non-renewable energy consumption in urbanisation, liberalisation, and economic growth in selected Asian developing countries. The estimation consequences identified population, non-renewable energy, energy consumption, and affluence as the critical factors in pollutant releases in Asian countries. Urbanisation is taking place is an unprecedented speed and scale in developing countries. It took approximately 150 years for Europe's development rate to rise from 10 to 50% in most Asian developing countries. That same shift is happening in a time that is one-third as long. The growth of urbanisation in Asia and Africa is much more rapid than in other regions. Urbanisation has negative and positive environmental effects by enhancing carbon absorption in the atmosphere, according to Chester. It induces high usage of energy and burning of fossil fuel through fast industrialisation, transportation of supplies and foods, and mechanisation of agricultural processes (Chester, Jones cited in Zhu et al. 2021)

5. Current Challenges of Energy and Energy Utilisation

There is a growing energy demand, and the obstacles of scarcity and environmental influence associated with traditional sources are at the base of a very likely energy crisis in the next two or three decades (Current Challenges In Energy, Openmind" 2022). We are facing various challenges in fulfilling energy and sustainable development needs. The two key challenges faced by the world first is how to handle the energy scarcity, and second is how to create and utilise energy to bring sustainability. Challenges to energy and energy utilisation are discussed here:

5.1. Finding New Alternatives

Finding newer alternative energy sources is the need of today's world. Orthodox energy sources are kerosene, wind energy, thermal power generators, and petrol Sun, is an enormous green energy source. Notably, terrestrial soil is exposed to a massive amount of solar energy, about ten thousand times all energy used worldwide. Energy utilisation is significantly affected by the laws of thermodynamics. The first law of thermodynamics declares that 'energy cannot be formed or destroyed, but it can merely transform into one form to another (Tripathi and Padmanaban, 2021).

5.2. Utilise Alternative Ways of Energy

There are unusual changes in the weather conditions due to human-induced climate. Change in climatic conditions is the primary focus for policymakers and businesses around the world. Generally, heavy transport duty and industries are resulting in greenhouse gas emissions. In India, there is an increasing growth in the industrial sectors, for example, cement and steel. The challenge is to utilise alternate methods and techniques of energy so that other harmful chemicals like Co2 can be reduced. Green hydrogen is one of the substitutes for carbon usage.

5.3. To Modify the Current Energy System

The rates of petroleum will rise continuously. The climate effects of massive usage of fossil fuels, current nuclear installations, and energy demand make it difficult to unrestraint any current sources. Therefore, there is a need for essential changes to decrease the environmental impact, and a new alternative source or green sources must be included (Zhang, 2017). India is focusing on reducing emissions by 2070, and India's central electricity authority (CEA) has done an energy mix estimate for 2030. The second goal is to source 50% of energy requirement from renewables. The third target is to reduce projected emissions from the current CO_2 levels. India attained a 25% emission power reduction in GDP between 2005 and 2016 and reached more than 40% by 2030 (Subhramanyam, 2022).

5.4. Increasing Urbanisation

Countries should fix the energy usage according to the level of urbanisation. It implies that countries need to use an enormous amount of energy in the early stage of urbanisation. Nevertheless, it will increase energy intensity after urbanisation reaches a specific level. In this stage, the energy efficiency in various countries will improve the fast development of urbanisation and enhance energy intensity reduction. Thus, it is lucid to determine energy usage following urbanisation (Zhu et al., 2021). Secondly, countries at various urbanisation levels in developing and developed countries have alterations in energy intensity, according to Li et al. 2020. Most of the OECD countries are created, and the rate of urbanisation reached about 80%, and urbanisation in these countries has come to a stable development stage. Thus, it does not need to enhance the urbanisation level at the price of a massive quantity of energy.

If developed countries devote much energy to urbanisation, it may slow economic growth and increase the greenhouse effect. However, on the other hand, they should protect the environment and control the greenhouse effect (Zhu et al., 2021).

Reduction of energy uses, and pollutants emissions will be likely along with urbanization if governments of these countries do the following:

- Support renewable energy development.
- Boost construction of renewable energy generation and supplying infrastructure.
- Develop a high energy-efficient and reducing industry-based emissions.
- Boost a free trade regime for clean technology from developed countries and promote urbanisation with low carbon urban infrastructure and transportation systems to gain sustainable growth in emerging Asian economies (Salim, Rafiq and Shafiei, 2022).
- The integrated energy system can be an extra to an current energy source (a hybrid solution) to decrease fossil-fuel consumption or a stand-alone for complete fossil fuel displacement.

- A pre-engineering study is behaviour determining economic and technical possibilities of a given system need and providing savings in the integrated energy systems, costs, and design time to plan the best resolution ("Integrated Energy Systems" n.d.).

5.5. Usage of Natural Resources for Power Generation and Energy Application

Due to eco-friendly features, energy utilisation is more natural in many energy domains such as power generation, energy resource, cooling, heating transport, and energy storage. However, it is not easy to use natural resources for energy implementations. Although natural resources use as fuels, working fluids, and refrigerants for an extended period, some problems still need to be solved. There is a need to reduce the adverse impact of energy utilisation on nature and human society. It should be feasible, efficient, sustainable, safe, and stable (Zhang, 2017).

Empirical studies concluded that the critical determinants of CO_2 releases in Asian countries are affluence, population, and non-renewable energy usage. The primary variable that is urbanisation tends to raise emissions though the consequence is not statistically significant. The results from linear and non-linear models expose that population and non-renewable energy consumption rise energy intensity while economic growth and trade openness tend to reduce energy intensity. The result demonstrates that population, non-renewable energy consumption, and affluence positively and significantly affect energy intensity. These results have some considerable policy implications. (Salim, Rafiq and Shafiei, 2022). In addition, fossil fuel energy systems are vital contributors to account for two-thirds of global GHG emissions and climate change. Moreover, fossil fuels harm energy security, and these challenges indicate a requirement to move towards green energy agenda for emerging and industrialised economies (Global Green Growth Institute, 2017).

Energy is one of the strategic issues of GGGI, and the energy sector plays a leading role in implementing and promoting green growth. Submission of NDC, including more than 180 countries preceding the Paris Agreement, represents the dominant role of the energy sector towards green growth. In addition, all INDC submissions comprise energy-related action, and around 50% of NDCs include specific energy-related targets to fight climate change (Global Green Growth Institute, 2017).

GGGI's significant activities in energy service delivery are to achieve energy outcomes. Accordingly, GGGI will perform to achieve its strategic energy outcomes with the following essential activities:

- Support green energy concept in the subsector plan and policies at national and subnational levels/energy sector.
- Support the realisation of green energy plans and policies with the development and financial structuring of financial instruments, financing vehicles, and bankable projects.
- Enhance multi-directional information, knowledge sharing, and learning in green energy development (Global Green Growth Institute, 2017).

Due to the increasing cost of electrical energy, thermal storage technology has been developed and, for example, applied to refrigerators. Global energy development is connected with resource endowment, and the earth is gifted with plenty of fossil energy, oil, coal, and natural gas. And also with large quantities of renewable clean energy such as wind, solar power, and hydro. However, global energy development has conventionally depended extensively on fossil energy, resulting in various problems like climate change, resource constraints, and environmental pollution. Thus, to avoid problems, humanity must adopt new features of globalisation on the economic development, environmental fronts, and resource allocation to set the stage for clean, efficient, secure, and sustainable development of world energy (Liu, Zhenya, 2015).

The integrated energy system is custom-designed arranging of diverse renewable technologies with concentrated solar power (CSP) turning into a focal point of the system. Several energy flows, for example, electricity, heating, steam, desalination, and cooling, can be provided by one carbon-free system, all periodically balanced to get the lowest likely energy costs. The energy plant is, thus, perfect for industries with dynamic energy requirements or costly access to energy, for example, greenhouse, paper industry, textile industry, aluminium industry, remote mining operations, breweries, improved oil recovery (EOR), etc. ("Integrated Energy Systems").

Many energy flows (heating, steam, desalination, electricity, cooling) can be provided with one carbon-free system, all seasonally balanced, to attain the lowest possible energy costs. The energy plant is thus ideal for industries with dynamic energy requirements and unreliable or expensive access to energy.

Global energy development is dealing with severe challenges regarding efficiency, allocation, environment, and resources, primarily because of environmental problems such as air pollution, climate change, and resource depletion resulting from large-scale fossil fuel utilisation. Other issues include low efficiency, high costs, and difficulties in energy allocation over long distances. Therefore, intensive efforts are required to encourage an energy revolution to meet challenges. Therefore, it is essential to drive the global progress of energy resources that are efficient, safe, and sustainable (Liu, Zhenya, 2015).

Researchers studied Nigeria's national energy outlook and found inefficient energy utilisation. Moreover, it revealed various energy conservation opportunities. Energy utilisation of the country was investigated, and the potential areas of energy conservation were identified in the major economic sectors: residential buildings, industry, transportation, and offices. Application of energy conservation actions includes manufacturing, residential buildings, transportation, conservation through waste control, power generation, offices, etc. Energy efficiency means the energy use in such a way that it will minimise the quantity of energy necessary to provide services. In Nigeria, much energy is wasted due to public and private offices, households, and industries using it more than required. It is due to inefficient and old methods and practices (Oyedepo, Sunday 2012).

Many renewable energy technologies comprise hybrid and related technologies. These are sued for storing energy produced through renewable energy, assisting in efficient energy delivery, and predicting renewable energy supply.

A green energy certificate can purchase to support green practices. It is estimated that more than 35 million European and 1 million American homes use green certificates. However, due to increasing temperature, agriculture output will fall, storms will increase, and damage from floods and disease. Therefore, the environmental cost is more, and loss is irreversible. Furthermore, due to growing temperature, flora and fauna will suffer directly from high temperature and indirectly through the harm to their habitats (Laxmi, Dudi, and Shika, 2015).

Conclusion

Energy utilisation and safeguarding the environment have increased attention over the last few decades. The issues discussed in implementing green energy and technology are finding new alternatives, utilising alternative ways of energy, modifying the current energy system, urbanisation and sustainability, and using natural resources for energy generation and applications. The energy utilisation system should be such that utilisation is more and wastage is less, reducing adverse environmental impact. Green energy is a new concept that has been reinvented to prevent the environment from being damaged. There is increasing urbanisation and demand for energy; therefore, the government should also focus on supporting renewable energy systems, developing highly energy-efficient, and decreasing industry-based emissions. Moreover, it boosts the construction of renewable energy generation and supplying infrastructure, using clean technology and integrated energy system.

Scarcity of access to energy affects economic, social, and environmental development and quality of life. Developments in the standard of living comprise increased agricultural output, industrial output and efficient transportation, health care, adequate shelter, and other human services, all of which need access to energy. Thus, establishing sustainable energy is a significant need for the sustainable development of the economy and society.

It is not easy to use natural resources for energy applications. Although natural resources have been used as fuels, working fluids, and refrigerants for an extended period, some problems still need to be solved. There is a need to reduce the adverse impact of energy utilisation on nature and human society. It should be feasible, efficient, eco-sustainability, safe, and stable. (Zhang 2017). Green energy can be used in residential or commercial spaces. The most common form of green energy is solar energy. The use of green energy is depended on the location. Places where the wind is in abundance, may rise wind turbines to generate renewable energy, which can be used for pumping water or charging sailboat batteries.

Biomass is a very available renewable energy source, and it is used for producing electricity and also used for transportation fuel. The usage of biomass as a renewable form is commonly known as bio-energy. Geothermal taps internal earth heat can be used for heating, building cooling, and electric power generation. Marine energy comes from various sources, including tidal and energy generated from ocean waves (Subhramanyam, S., 2022). Wind farms are seen as the most efficient energy source as they need less processing and refining energy than the production of solar panels—advanced technology-assisted to improve lifespan and thus LEC of the wind turbine. Green energy solutions do not require additional energy expenditure (TWI, n.d.).

Chapter 2

Integrated Energy Systems

Abstract

Climate change and energy crisis are the two main issues discussed worldwide. An integrated energy system is one of the steps to having a steady supply of renewable energy. The principal goal of integrating the energy and consumer sectors is to decarbonise the global economy to a large extent. It is new hope for the future sustainable energy system. It is the procedure of coordinating and directing the planning and operation of an energy system across several pathways or geographical scales to bring cost-effective energy with less impact on the environment. It is flexible and permits accessible energy to other areas. Therefore, it is necessary to focus on the integrated energy system, sector integration, integrated energy policies, integrated energy network, clean energy, and technologies to develop sustainable energy. Therefore, this chapter focus on the integrated energy system, which is the future of a sustainable energy system.

Keywords: Energy system integration, renewable energy, clean energy, integrated energy network

1. Introduction

Integrating the energy system spanning heating, electricity, transportation, and cooling gas is an effective strategy for increasing energy system elasticity and balancing energy source volatility. In addition, it contributes to the creation of an integrated energy system that is energy efficient ("Integrated Energy Systems - Home," n.d.).

Renewable energy's share of total electricity generation is steadily increasing. According to the Renewable Energy Policy Network for the 21st century Renewables 2018 Worldwide Status Report, renewable energy accounted for approximately 20% of global final energy consumption in 2016.

According to data from the Fraunhofer Institute for Solar Energy Systems, 41% of net power generation (i.e., the electricity that comes from the socket and is used in a business or house) in Germany was renewable in the first half of 2018. As a result, wind and P V systems significantly contribute to the global energy supply. The disadvantage is that solar and wind energy are inherently inconsistent, as they create varying amounts of energy. In the future, the sun or wind could reach peaks in the generation that surpass demand. Therefore, renewable power plants must be limited in the future. In addition to electricity, there are various forms of energy. However, energy is still obtained mainly by burning fossil fuels for transportation and heating. According to the REN21 renewables 2018 Global Status Report, renewables account for 10% of total energy consumption, whereas the transportation sector accounts for only 3%. Together, these two sectors consume more than 80% of total energy (48% for heating and cooling and 32% for transportation). Thus, they are also accountable for a large proportion of energy-related CO_2 emissions (Baars 2018).

Battery storage systems and power-to-X technologies flexibly transform surplus renewable electricity into various forms of energy, increasing the role of renewables in the transportation and heating sectors. The "X" represents the most likely energy carriers. The most common examples of these technologies are heat, gas, and battery storage systems. A viable energy system is made possible by integrated energy, and it is on the rise. Renewable energies accounted for 20% of global energy consumption in 2016, according to the REN21 renewable 2018 global status report. According to the Fraunhofer Institute for Solar Energy Systems, renewable energy accounted for 41% of Germany's electricity output in the first half of 2018. P.V. and wind energy thus contribute significantly to the global energy supply. The disadvantage is that wind and solar energy are fickle since they create varying amounts of energy. In the future, too much wind or solar could lead to a peak in energy production that exceeds demand.

There is a limited number of power plants and a loss of electricity. Integrated energy permits energy to be transferred from one sector to another as there is more to energy than electricity. Integrated energy connects the energy-related sectors of heat, power, transportation, and cooling to allow for the use of a substantial proportion of renewable energy sources such as wind, solar, and biomass. Extensive use of natural gas or electricity refuelled with natural resources is helpful in transportation. Pumps and heating elements, for example, are valued in the field of heat, and they use renewable electricity to provide heat for domestic consumption (Baars 2018).

Integrated energy makes it possible to have a steady supply of renewable energy. Furthermore, in cities, there are natural sources of municipal waste. For example, food waste can transform into beneficial products, including biogas. Furthermore, these can be used for heating, power, or transportation (Baars 2018).

2. Energy System Integration

Energy system integration coordinates and directs energy systems' planning and operation across several pathways or geographical scales to bring cost-effective energy with negligible environmental impact and deliver reliable services.

Figure 3. Air pollution.
Source: https://cdn.pixabay.com/photo/2016/11/29/02/15/air-pollution-1866788_960_720.jpg.

Energy systems have progressed from individual systems with little or no dependencies into a complex set of integrated systems at scales that comprise cities, regions, and customers. This progression has been determined by economic, political, and environmental objectives. Flexible energy systems are needed to meet the globally identified carbon emissions targets by deploying sizeable renewable energy capacities while maintaining reliability and competitiveness. The different integrated system is to provide flexibility.

Physically connecting energy vectors (thermal, electricity, and fuels) allows for an integrated energy system.

Urban energy resources have a low carbon footprint and can decrease carbon emissions. It is valid for countries where most of their production comes from fossil fuels and Industries increase air pollution, which harms the environment (Figure 3). Thus, local resources show an opportunity to reduce dependency on external resources and to release net capacity and electricity for other needs.

3. Development of Integrated Energy System

Interconnections between energy vectors such as heat, electricity, and fuels, as well as interactions with other large-scale infrastructures such as transportation, water, data, and communications networks, are all part of ESI. ESI is particularly useful at institutional, physical, and spatial interfaces, where fresh opportunities and difficulties for demonstration, study, and application to reap its societal and commercial benefits abound. Engineering, science, technology, economics, regulation, policy, and human behaviour all fall under the umbrella of ESI.

An integrated energy system consists of multiple energy inputs (such as fossil with carbon capture, renewable, nuclear), multiple energy storage choices (electrical, chemical, or thermal), and multiple energy users (such as industrial heat, transportation fuel users, electricity users, and grid consumers) (Baars 2018).

3.1. Meaning of Integrated Energy System

Power plants provide energy, but they also generate much heat. Today, 40% of all energy is wasted because it cannot use for other activities requiring thermal energy. Therefore, the plant owner and the environment would benefit from using efficient power.

An integrated energy system is a power plant that produces both heat and electricity. It is a system that could bring together economic growth, resource efficiency, competitiveness, and job creation.

The U.S. Section of Energy's Office of Nuclear Energy supports a national laboratory integrated energy systems (IES) effort. This initiative conducts research and development to expand nuclear energy's role beyond assisting the electrical system. One of the tasks is providing power to diverse industrial, energy storage, and transportation uses ("Integrated Energy Systems - Home," n.d.).

3.2. Reason of Integrated Energy System Development

Integrated energy is a critical technology in energy change. It connects the heat, individual electricity, heat, and mobility energy sectors, ensuring that renewable energy is used effectively. The only means to decarbonise the world economy is through this strategy. An integrated energy system is the most efficient way of supplying energy in the digital age 4.0.

The function of integrated energy in the energy transition is critical. It combines heat, personal power, and transportation energy sectors, proving that renewables may use efficiently. It is the method for decarbonising the world economy. Integrated energy and digitisation are finding the way for the future of energy supply and opening up new doors (Baars 2018).

The significance of Energy Systems Integration (ESI) is in directing how to develop energy systems and transport energy in various forms at appropriate scales to achieve economic, dependable, or environmental goals. Industry and government may learn about the best methods and techniques for accomplishing these goals by studying and designing integrated energy systems (O'Malley et al., 2016).

It is simple to categorise ESI into a few categories because it incorporates all types of energy sources and end-use applications. The three categories of ESI are simplified, synergise, empower, and opportunity. These categories aid in identifying how different ESI techniques can handle issues that have risen to the top of global energy agendas. Because ESI is such a broad concept encompassing a wide range of energy sources and end-use applications, it is easy to group instances into a few categories. The three categories of ESI scenarios are: simplify, synergise, empower, and opportunity. These categories help distinguish how different ESI strategies can deal with problems (O'Malley et al., 2016).

3.2.1. Streamline Areas
The reform, restructuring, and upgrading of the current energy system at institutional levels such as infrastructure investment, laws, markets, and regulations are all covered in this topic. In addition, enhancing energy end-use

flexibility is expected to offer system-wide benefits and the potential to create new product and service markets.

However, we will need appropriate regulatory and market structures, physical energy network features, new operational and planning paradigms, flexible end-use products, and an integrated communications system to reap these benefits. Currently, such a system does not exist, necessitating a system-wide understanding to deliver pragmatic and long-term solutions. In addition, developing more integrated energy system-wide policies will allow for better risk management. Therefore, capacity and flexibility of a more integrated energy system are helpful. (O'Malley et al., 2016).

Furthermore, by removing the institutional barrier between distribution and transmission networks, distributed resources will be better integrated, and regional integration will improve. The energy sector will rationalise if it provides standardised requirements, updated interconnection and interoperability standards, and grid codes. The flexibility of integrated energy systems will improve by investing in appropriate infrastructure.

The synergy between energy systems labels ESI solutions that connect energy systems across energy domains and spatial scales to gain efficiency and performance benefits. Synergies explain ESI solutions that link energy systems between energy domains and across spatial scales to take benefits in performance and efficiency. To date, the connection of the electricity and heat sectors has been aimed at the supply side, for example, combined heat and power (CHP), for fuel-saving purposes. However, its inflexibility can lead to sub-optimal whole system performance at the system level. A good instance of this is wind curtailment in China, partly due to the incapacity of physically inflexible CHP plants to decrease electricity generation while providing heat. For example, ESI solutions that combine heat storage into the CHP plant are being developed and show a shift from the supply to the demand side, for example, electrical heating of water, heat pumps, and thermal storage in buffers. It is probable to capitalise on "vital storage" where the flexibility in one part of the system, transport, heat, water, etc., can be integrated with, for example, the electricity system and used similarly to electricity storage. This virtual storage can be significantly cheaper than dedicated storage as it does not need significant capital investment but needs a more integrated energy system (O'Malley et al. 2016).

Demand management technologies, such as heating, controlling, and cooling loads, are currently being developed and deployed partly by leveraging this virtual storage. Empower means ESI actions that compromise the consumer, whether through their investment decisions, their active involvement, or their decision to shift energy modes. Investment in energy efficiency is increasingly a cost-effective way to decrease energy demand. It can lead to a system-wide advantage that comprises upstream capital and operational savings. Energy efficiency enhancements or targets also contribute to broader policy and social goals, notably macroeconomic efficiency, public budget balance, industrial productivity, security of supply, and health benefits. The issues raised about how best to integrate thermal grids into the electricity grid significantly impact how consumer needs will meet and whether consumers will accept them. Consumers can also make selections that provide the required services while using less energy by mode shifting. For example, consumers can choose to take public transport in place of their vehicles. ESI is a significant notion of making the energy system more flexible, enabling the efficient combination of renewable energy and decreasing carbon emissions. It ranges from very simple to very complex. Therefore, it is necessary to understand the ESI value proposition and communicate it to inform and educate professionals (O'Malley et al., 2016).

3.2.2. Flexible Energy
Flexible energy permits the available power to be moved to other areas. Integrated energy links the energy-related sectors of heat, electricity, cooling, and transport so that renewable energy from solar, wind, or biomass can be utilised as much as possible. In transportation, extensive use of natural or electric gas vehicles refuelled with natural resources is of noticeable benefit. In the area of heat, for instance, great significance is being attached to heat pumps and heating components. They produce heat for household use through renewable energy and generate heat for household use through renewable electricity. Due to synergy impacts between heat and electricity, energy can be put to good use precisely when it is required (Baars 2018).

3.2.3. Integrated Energy Assist the Global Climate
The overarching goal of integrating the energy and consumer sectors is to decarbonise the global economy to a large extent. In December 2015, the international community agreed in Paris to limit global mean temperature rise to no more than two degrees Celsius above pre-industrial levels (COP21). Under international law, the agreement approved in November 2016 is legally binding (Baars 2018).

3.2.4. Integrated Energy Makes Possible Sustainable Energy

3.2.4.1. Generating Gas from Electricity: Power-to-Gas (P2G)
The current gas network architecture can already store and deliver massive energy, and this storage can also utilise renewable energy using power-to-gas technology. For example, this allows CO_2 released during biogas production to be used to produce methane, which is then available in the natural gas system as a raw material for the chemical sector, as well as propulsion energy for aircraft and cars, or re-use at gas-fired power plants. Integrated energy for the home: By combining a heat pump, solar system, battery storage, and intelligent energy management, a household's energy supply can be more efficient and sustainable (Baars 2018).

3.2.4.2. Generating Heat from Electricity: Power-to-Heat (P2H)
Excess electricity from renewable energy sources can also be utilised to supply heat. Renewable energy provides heat surplus electricity, obtained as heating energy or for making hot water. Power-to-heat devices are hybrid systems with a heat producer operated by a traditional fuel like natural gas or wood. In the event of surplus electricity, heat is accessible from electrical energy. Then the conventional heating system is used. The heat can be fed into the district or local heating grid or used to supply heat to separate buildings or industrial companies. The heat can also be temporarily stored using an extra buffer tank and recovered as needed to provide negative balancing energy (Baars 2018).

3.2.4.3. Storing Electricity in the Batteries
Battery storage systems rely on local renewable energy surpluses, and rechargeable chemical cells (battery accumulators) absorb and release energy more quickly in this situation. Battery storage systems depend on local renewable energy surpluses, and rechargeable chemical cells (battery accumulators) consume and release energy more rapidly in this situation. In addition, battery storage systems coupled with Photovoltaic systems (P.V. systems) demonstrate greater energy productivity and provide an electricity supply during grid disturbances in the home storage zone. The batteries provide an operating reserve in large storage power plants up to the M.W. range. Furthermore, battery inverters are now involved in fast frequency stabilisation and quick reserve (coverage of variations due to inertia). It is important when extra power is needed because there is always a large load somewhere that requires much energy in the short term.

For example, the soccer stadium's floodlight system shifts on the generator rotors' momentum, delivering stored energy and stabilising the electricity grid at short notice. If battery inverters substitute the rotating masses of power plants, this provides another vital advantage. A traditional power plant can only ever offer a fraction of its power as a functioning reserve, and a battery storage system can offer its full nominal power. For instance, a battery storage system of 30 megawatts substitutes the operating reserve capacity of 1,000 megawatts (1 gigawatt) of power output (Baars 2018).

Even in the event of widespread grid failures, Battery investors can confirm that the grid will soon be accessible again. They can black start, initiating up independently of the electricity grid and securing the electricity supply (Baars 2018).

3.2.5. Integrated Energy Confirms Grid Quality
Reliable supply: renewable power plants offer standby power and confirm grid stability (Baars 2018).

Just as renewable energy production varies, energy demand is not always similar. Therefore, the energy balancing market reimburses for fluctuations between production and consumption in the utility grid. To confirm grid quality is maintained at all times, power plants must make balanced energy accessible where required within seconds (Baars 2018).

If demand exceeds power generation capacity, electricity must be speedily fed into the grid (positive balancing energy). If, however, the electricity supply is greater than demand, the electricity must take off the grid (negative balancing energy). Therefore, power plant operators in the balancing energy market obtain a feed-in tariff for their balancing energy (Baars 2018).

3.2.6. Storage for Cases of Need
A fascinating possibility is large battery storage systems or power-to-heat systems. Both technologies can make capacity available or store surplus energy more quickly than fossil fuel power facilities. Battery storage systems on the grid, for example, have an energy backup that they may provide as needed. On the other hand, biogas plants do not instantly feed the electricity; instead, they temporarily store it in the thermal energy storage system used in conjunction with the P2H system.

They provide energy when the wind is not blowing, or the sun is not shining, substituting demand for fossil fuel-run capacity and resulting in CO_2 emissions (Baars 2018).

3.2.7. Integrated Energy Is the Statement of Energy Freedom for Consumers

Integrated energy transforms passive electricity users into prosumers who can participate actively in the energy supply system. Instead of being a passive investor, integrated energy allows the individual consumer to shape the energy conversion actively. Numerous solar and smaller wind turbine systems with private operators or in public parks illustrate that public awareness of the need for a sustainable energy supply is growing. Renewable energy is appealing not simply because of financial incentives. Individual households and businesses are becoming more interested in novel technologies that allow them to become self-contained power supply firms and fossil fuel producers. Through intelligent energy management, P.V. system operators can coordinate consumption and production to use the largest probable share of their self-produced renewable electricity (Baars 2018).

At a similar time, decentralised power production plants and the digitalisation of energy conversion are turning P.V. system operators into "prosumers" who can contribute to the energy supply system through direct marketing. Until recently, that was the preserve of enormous power supply companies and public utilities. The operator can now also sell self-generated renewable electricity with the digitalisation system.

The likelihood ranges from today's direct marketing to future contributions in peer-to-peer or flexibility markets. In terms of technology, renewable production systems have even made it possible for private households, commercial enterprises, or industrial plants to set up their electricity supply independent of the electricity grid (Baars 2018).

3.2.8. Green Electricity versus Mere Theory

The theory needs to be followed by practical action. Though various renewable energy has occurred worldwide, we have shut down any fossil fuel power plants. So, what good is green electricity if nuclear power plants and fossil fuels continue to emit harmful gases or generate radioactive waste, endangering human and environmental health? Integrated energy demonstrates that fossil fuel power plants no longer need to be operated at base load to provide necessary balancing energy. With the elastic interconnection of the individual energy sectors, this is probably a way that is much more climate-friendly and cost-effective (Baars 2018).

Integrated energy is the only way to reduce CO_2 emissions and rescue the environment. Only an integrated renewable energy system can achieve energy change—flexible systems, the advancement of promising decentralised production technologies, and digitalisation all present promising opportunities. Integrating energy is a win-win situation for a reliable energy supply and a better environment (Baars 2018).

4. Sector Integration

Sector integration is a smart step toward a carbon-free economy. Sector integration will be crucial in decarbonising energy systems and reducing CO_2 emissions to fight climate change. The main factor in decarbonising the energy economy is not how much renewable energy can generate but how it can be integrated into the energy system. Heating, transport, cooling industry, and water treatment are all sectors whose demand is elastic enough to exploit renewable energy fully. Sector integration is a linking force across industries to exploit the potential of renewable energy. The key to sector integration applies to a system that can consume energy or deliver power to another sector. Various industries, even retail, produce heat as a waste product, which can then be oppressed elsewhere to create a more sustainable energy system. Sector integration is specifically pertinent when energy generation is based on renewable energy like sun and wind. Sector integration permits for electrification of more sectors and enhances required resilience as the demand for power does not always track the weather. Integration facilitates energy systems to use and reprocess energy more professionally. Example district heating system that can use electricity when it is abundant and inexpensive to heat water stored in tanks and pipe network to be used when electricity is expensive ("Integrated Energy Systems," n.d.).

4.1. Benefits of Sector Integration

Sector integration aids in the achievement of climate goals and the development of cost-effective technologies. The most challenging task is to create a carbon-free energy economy. When demand is insufficient to utilise the grid's capacity fully, power is generated. Sector coupling with thermal storage, which allows for flexible power consumption, evens out the supply and demand gap, allowing the grid's capacity to be fully utilised. Sector cooperation benefits both climate change aims and cost-effective systems. The fundamental goal is to create a carbon-free energy economy. Electricity demand is made when an order is insufficient to fully utilise the grid's capacity. Sector coupling and thermal storage can even out supply and demand imbalances, allowing the grid to be fully used.

Smart energy systems will enjoy cost-efficient operation as they adapt their power purchase to exploit the lowest possible rates at any given period. As a result, the power supply can avoid the loss of revenue and curtailed capacity. Furthermore, the principle of sector coupling assists in reducing costs, and this is due to synergies relating to the freezers in supermarkets, heating, cooling of buildings, battery charging, etc. The European Union accounts for half of the energy consumed, and 75% of it is based on fossil fuels.

The park's integrated energy system is critical to achieving carbon-neutral and long-term energy development. Its optimised dispatch can improve energy utilisation efficiency and lower energy system operational costs. ("Integrated Energy Systems," n.d.).

According to Zahedi, the park integrated energy system is one of the necessary forms of dispersed renewable energy utilisation. It plays a significant role in promoting the absorption of renewable energy and establishing a new power system under lower carbon (Zahedi cited in Li et al. 2021). The dispatch plan is updated according to the direction and coordination of renewable energy and load to ensure economic cooperation. Furthermore, reduce the volatility of renewable energy and load, enhance the feasibility of the dispatch plan according to Fan and create an integrated electricity gas thermal system planning model bearing in mind demand response, energy conversion, and energy storage. Ai and Jiang move forward with an integrated energy system representing an optimisation operation model. It is considering the novel energy incentive mechanism to simplify the influence of various incentive mechanisms on the economic benefits (Fan, Ai, Jiang cited in Li et al. 2021). Cen et al. established an integrated energy utilisation model and a structure virtual energy storage system under the TOU price. Lin and Fang found a local integrated energy management system model based on a multiagent method to optimise operator profit and user cost. A control optimisation approach is presented because of economic performance and response performance to progress the multi-energy optimisation of interconnected systems (Lin, Fang cited in Li et al. 2021).

Integration of renewable energy needs a robust transmission grid that can handle fluctuations from, for example, sun and wind energy. Integrating energy systems, gas, cooling, electricity, heating, transportation, and water is adequate. It enhances the flexibility of the energy system and assists in balancing the fluctuations of energy sources in a financially feasible manner ("State of Green: Connect. Inspire. Share. Think Denmark" 2018)

Evolving coordinated systems through ESI analysis needs a proper understanding of the various actors involved, along with their incentives, motivations, and the information and knowledge they have access to. The actors performing in each energy domain tend to perform on the information they have in a manner that maximises advantages for their domain but not for the whole energy system. For example, each user consumes energy based on their needs and requirements, each market values specific investment outcomes, and each government aids its political and social motivations. Still, there may be no coordination across these domains to fix the best option for all actors performing. A poorly executed energy transition could result in an energy system lacking social equity, technical integrity, or political acceptability. Integrated energy systems, sector coupling, and sector integration are various names. Still, the objective is to create an intelligent energy system that connects energy-consuming sectors to the power grid to optimise the synergy between energy production and its use (O'Malley et al., 2016).

5. Challenges of Integrated Energy System

In an integrated approach, future investment in generation capacity and grid infrastructure must compare to investment in district heating network development to achieve a cost-effective outcome. The current region heating networks have a high-temperature supply, and integrating low-temperature urban waste heat sources is challenging. Furthermore, local temperature networks in building areas can result in this challenge. Energy storage thus plays a significant role in assisting the demand at peak times. Proper management of the various energy resources becomes significant where different energy resources and carriers can interrelate in an integrated energy system. Understanding the challenges and taking steps toward more sustainable, efficient, and reliable energy systems is vital to developing new understanding through innovation and research with public and private players.

In the literature, Zhang and Wei (cited in Li et al. 2021) established a two-layer capacity distribution optimisation method for IES based on the joint power supply scheme, cooling, and heating. Pu et al. (cited in Li et al. 2021) proposed an ideal operation method for the power system to convert the energy supply mode of the port integrated energy system. Lv et al. (cited in Li et al. 2021) proposed an optimised operation model of IES and explained the flexibility of demand response under various coupling degrees. Zhao et al. (cited in Li et al. 2021) proposed an electric gas thermal system and rolling operation optimisation strategy because of energy demand response and wind penetration. Qi et al. (cited in Li et al. 2021) established a two-layer collaborative optimisation method of IES because of D.R. and E.S. Yu et al. (cited in Li et al. 2021) a multi-energy flow calculation model, ideal scheduling model for the gas-power heat cooling integrated energy system established because of the integrated demand-side response, energy storage and wind power output prediction. Wang C et al. (cited in Li et al. 2021) proposed a demand response model based on a compensation mechanism to describe the scheduling elasticity of various load types. Wang J et al. established a two-state optimal scheduling model for an integrated energy system (cited in Li et al. 2021).

According to Duan et al. (cited in Li et al. 2021), the time requirement of varying energy flows on balance between supply and demand. In this respect, Yang et al. (cited in Li et al. 2021) must study the response strategy and cost of load under varying time scales.

Focusing on IES development on improved utilisation of low or non-emitting energy generations options within IES will assist the U.S. in achieving the goals established by the Biden administration to achieve net-zero emissions by 2050 and a clean energy economy ("Integrated Energy Systems," n.d.).

Every system will approach ESI from a diverse starting point; for example, an urban area in the developed world will have a diverse approach compared to a rural area in developing countries. Therefore, it is vital to describe the geographical scope, the elements, the impact of the surroundings, and the boundaries. For instance, renewable integration is the driving force in various regions but not all. In some areas, the key drivers are raised combined power and heat, a shift from coal production to natural gas, increased efficiency, or electrification. Various incentives, access to capital due to location, and decision-making processes will result in diverse energy systems and approaches to ESI; for example, a government can finance high voltage transmission while individuals will not. Each energy develops, and it will be essential to monitor and reevaluate the system to judge how it is coordinated in the best manner (O'Malley et al., 2016).

6. Advanced Research on Integrated Energy Systems (ARIES)

ARIES is a research platform that can meet the complexity of the novel energy system and conduct integrated research to develop modern technologies ("ARIES: Advanced Research on Integrated Energy Systems," n.d.). ARIES represents a substantial scale-up in experimentation capability from current research ("ARIES: Advanced Research on Integrated Energy Systems," n.d.). It makes it possible to understand the impact and get the most value from the millions of new devices. For example, renewable production, energy storage, vehicles, hydrogen, and grid-interactive effectual buildings are connected to the grid daily ("ARIES: Advanced Research on Integrated Energy Systems," n.d.). The platform's scale will also make it probable to consider chances and risks with the growing interdependencies between the power system and other infrastructures like transportation, natural gas, telecommunications, and water ("ARIES: Advanced Research on Integrated Energy Systems," n.d.).

Aries unites research capabilities at multiple scales and across sectors to develop a platform for comprehending the total influence of energy system integration. ARIES point out the risk and chances of widescale integration across five research zone ("ARIES: Advanced Research on Integrated Energy Systems," n.d.).

ARIES links multiple individual energy storage applications with a system-level viewpoint. The link of at-scale storage technologies such as batteries +thermal or batteries +hydrogen will assist essential steps toward confirmative energy system models and controls. As storage technologies graduate from the laboratory to the multimegawatt level, ARIES will support systems to stay ahead of function and interfacing challenges linked with scaling ("ARIES: Advanced Research on Integrated Energy Systems," n.d.).

The continued progress in power electronics is developing a new paradigm in power system operation. ARIES assistance addresses the basic modifications between power electronic-based equipment and conventional devices and the restrictions overcome to increase renewable production.

Integrating new system architectures and electronic technologies, ARIES will assist a future grid with resilient and elastic operations ("ARIES: Advanced Research on Integrated Energy Systems," n.d.).

6.1. Hybrid Energy Systems

With future energy systems predictable to include millions of distributed energy assets, the ARIES research platform is exclusively able to replicate the diverse time scales, technologies, and physical scales. ARIES announces an immediate real-world environment with real-time models, physics-based, high-fidelity that facilitate the links between hundreds of actual hardware devices and tens of millions of replicated devices. This research will advance the basic science for real-time optimisation and control of large-scale energy systems ("ARIES: Advanced Research on Integrated Energy Systems," n.d.).

6.2. Future Energy Infrastructure

ARIES assists the innovation required for following production energy infrastructure solutions. The future energy infrastructure research concerns communication and delivery networks for diverse advanced fuel kinds and infrastructures that undergird the power, buildings, transportation, industrial, and manufacturing sectors. ARIES will make possible testing on grid designs that span microgrids up to high-voltage straight present transmission and communication grids. Furthermore, management and control systems optimally mix power delivery for various fuel and technology kinds ("ARIES: Advanced Research on Integrated Energy Systems," n.d.).

6.3. Cybersecurity

ARIES assist closes the system-level security breaks that arise from dissimilar hardware and software becoming integrated. The ARIES platform comprises monitoring, visualisation, and data processing for ARIES research assets and their links. By developing a digital twin of clusters of research hardware, ARIES could simulate and find out attacks on communication and control systems that are still growing with an effect of decreasing overall weakness in energy systems ("ARIES: Advanced Research on Integrated Energy Systems," n.d.).

This article adopts science and technology education methods and frameworks for investigating energy system integration (ESI).

ESI, the integrated operation and planning of various energy demands and supplies, can participate in the enhancement of energy reliability, flexibility, and security and thus enable a change to a low carbon economy. The article investigates U.K. research towards united computer models of energy system vectors. The research is built on fieldwork and involves a large ESI research project in the U.K., drawing on interviews, observations, and workshops.

Three main participants in the ESI and science and technology studies (STS) research of modelling are; ESI researchers are conscious of the economic, political, and social context of their work, though various of these frameworks are problematic to integrate with any valuable modelling process. Second, issues relating to policy and science around energy systems motivated ESI investigators yet were also understated in project work.

Third, to grow unique mixed ethnographic methods for study ESI permit researchers to build faith relations with research participants. Fourth, it discusses sensitive topics, such as the ethics and politics of energy modelling approaches ("ARIES: Advanced Research on Integrated Energy Systems," n.d.).

ESI means the integrated planning and operations of many energy demands and supplies such as electricity supply, natural gas, transportation, heating and electricity, energy network, etc.

Various studies do not investigate the associations between ESI research practices and the normative inference of energy integration in the making. There are the following questions that need to answer to discover how integration is driving energy system research and development:

- How do energy system model researchers theorise energy integration?
- How are system models supposed to support communication between science and policy?
- How can ethnographic research methodologies in STS report these matters?

Silva et al. studied based on fieldwork and were involved in a large ESI research project in the U.K., drawing on workshops, interviews, and observations (Silvast, Abram, and Copeland 2021). This study starts by providing a background on the arrival of ESI, a context for examining the emergency of ESI, a framework for analysing the representation of ESI, and materials and methods organised in this research.

Energy Systems Integration (ESI) coordinates coordinating the planning and operation of energy systems across several pathways and geographical scales to deliver cost-effective energy services and provide reliable and minimal environmental impact (O Malley et al. 2016). Britain's continued involvement in global energy infrastructure and markets was taken as a given despite the context of Brexit. On this foundation, an additional detailed discussion of modelling schemes needs to understand (Silvast, Abram, and Copeland 2021).

Conclusion

The global energy site will change more in the next ten years, and the world energy sector will move from fossil fuel to renewable energy sources. Industrial organisations are challenged with addressing this evolution in a transformative mode. Digitalisation will be crucial to making power-producing assets more efficient, the electric grid more resilient and secure, assisting manufacturers in reducing waste, and the aviation industry more sustainable ("Digital in the Future of Energy | G.E. Digital," n.d.).

An integrated energy network way leads to a more interlinked and integrated energy system that provides customers with improved selection and control, the reliability, affordability, and reliability they suppose, and the extra benefits of a cleaner and electrified environment. The integration of energy sectors provides opportunities to absorb surplus electrical energy from renewable energy sources, store it and provide backup supply during periods of high demand and increased prices, thus helping as an additional source of security and flexibility in energy systems. However, this needs the development of innovative optimisation and control solutions and suitable holistic planning strategies. An integrated energy network focuses on the main topics ("Home," n.d.). This chapter discussed integrated energy systems and methods. Several questions require answers, like the most significant research and development needs? How do energy systems and infrastructure requirements develop to provide increased value to customers and society? What are the chief implications for regulations, policy, and the creation of standards? Furthermore, how can the energy sector direct and focus investment currently, and what roles can electrification and electricity play in a more integrated energy future, etc.?

According to Dr. Aidan Rhodes, the most energy system is interdependent but not integrated. Variable renewable electricity and lower-carbon fuels provide energy services traditionally provided by higher-carbon sources. It might be realised by a more resilient system comprising an integrated energy system. The fundamental problem would be the complexity of a highly integrated system. It would be worsened by the existing fragmented nature of institutions and market structures in different energy sectors. There are two important hurdles, the first is the large requirement for multidisciplinary research, and the second is development efforts to integrate disparate technologies and systems fully. The second obstacle is the requirement for substantial policy assistance to make the integration. Third, there is a requirement to change market arrangements to increase investment in new and diverse types of flexibility—likely areas for integration between various energy systems. Finally, there is a need for long-term visionary political thinking and to think about what kind of energy is needed in the 21st century (O'Neil, 2018).

Clean energy has been rousing researchers to identify solutions to global climate change, specifically in utilisation and carbon capture. Carbon dioxide has been used for decades with technologies in various industrial procedures such as CO_2-enhanced oil recovery, food industry, urea production, the production of fire retardants, and water treatment. There are new co2 utilisation technologies at different phases of commercialisation and development. These technologies provide opportunities for emission savings for industrial sectors and power by partly replacing fossil-fuel raw materials, using renewable energy, and producing revenues and renewable energy through producing marketable products (Extavour, 2021). An integrated energy system is one step that focuses on utilising energy efficiently. It has two key benefits; first, it is a cost-effective way of energy, and second most important reduce the environmental impact. It plays a significant role in sustainable energy generation and less environmental damage. It has various benefits, such as negligible environmental impact, reliability of energy supply, lower electricity charges for consumers, etc. (Baars 2018). It makes it possible to have a stable supply of renewable energy. Energy system integration is coordinating, coordinating, and directing the planning and operation of energy systems across various pathways or geographical scales to bring cost-effective energy with less impact on the environment and deliver reliable services. An integrated energy system brings elastic energy, which allows accessible energy to be moved to other areas. Much energy is wasteful if it cannot use in different areas. Generating heat from electricity and gas from electricity allows energy to be transformed from one form to another. Sector integration is a smart step taken to make a carbon-free economy. It reduces carbon emissions to fight climate change. Transport, heating, cooling, industry, and water treatment are all sectors whose demand is flexible enough to exploit renewable energy fully. It works as a linking force across sectors to exploit renewable energy. In an integrated approach, future investment in production capacity and grid infrastructure needs to be compared to investment in district heating network development to attain a cost-effective outcome. Climate change and energy crises are one of the most significant issues in front of the world. It is necessary to overcome this problem. Thus, to resolve the issue of energy integration system, sector integration, integrated energy network, and clean energy system plays a significant role.

Chapter 3

Green Energy

Abstract

There is an increasing demand for energy globally, and the future is moving towards finding sustainable and environmentally friendly energy. Green energy is produced from natural resources, for example, sunlight, water, or wind. The objective of green energy is to protect the environmental system and meet the current and future energy demands, which leads to improving situations such as economic, social, human, and environmental changes in technologies. Dependence on conventional energy sources, increasing pollution, increasing demand for energy, and emission of harmful gases due to energy generation are the problems we face today. Green energy is a solution to all these problems; thus, it is necessary to focus on green energy. There are various hurdles to adopting green or renewable energy, such as cost, policy, and market share barriers. However, this green energy is the need of today's world. It is a sustainable global energy system that limits emissions and optimises energy efficiency. The global economy and technology must progress in harmony with sustainable development. This chapter discusses various types of green energy and highlights green energy, which promotes sustainability and is environment friendly. In addition, it also deliberates hurdles in the journey of the green energy system.

Keywords: green energy, clean energy, renewable energy, sustainability

1. Introduction

The word 'Green' prepare our mind to think about a world in an eco-friendly manner. Thus, green energy reflects the notion that the generation of energy from natural resources like wind, tides, sunlight, rain, plant, algae, geothermal heat, etc., have no or less impact on the environment and can be transformed (Shah 2021).

This energy is helpful for the planet, people and animals. Green energy can lead to stable energy prices as these sources are generated locally and not affected by geopolitical crises, price spikes, or supply chain disruption. The economic benefits also include job creation in building the services that mostly serve the communities where the workers are employed. Due to the natural energy generation through sources like solar, and wind power, the energy infrastructure is more flexible and less dependent on centralised sources. It is also a low-cost energy-saving solution (TWI, n.d.).

Renewable energy is vital for the energy system in the future and supplies urgent needs for its sustainable development, environmental impact, and usage. Due to the current problems of energy and environmental problems, it is urged the development and movement of renewable energies (Salvarli and Salvarli 2020). The development of renewable energy-dependent on technology innovation and enhancement of new high technology levels that fit industrialisation and commercialisation. The reality is that the cost of renewable energy development is somewhat high. If the government's assistance and policy presentation cannot confirm that a large-scale development country will not aid in reducing cost, maintaining reliability, increasing profit, and enhancing the value of renewable energy (Salvarli and Salvarli 2020)

2. Green Energy

The notion of green energy began in November 2006 as a renewable energy standard offer program. That offer was, in short, known as a source of power (SOP) or Renewable Energy Standard Offer Program (RESOP). It presented 20-year feed-in tariffs for wind, hydro, solar (PV), and biomass projects. The Ontario Green Energy Act (GEA), formally the green energy and green energy economy Act 2009, was presented in the Ontario Legislature on February 23, 2009, to enlarge renewable energy generation and encourage energy conservation (Shah 2021).

Green energy generates from many natural resources, such as sunlight, water, and wind. It recurrently comes from renewable energy sources. The energy does not harm the environment through factors such as releasing greenhouse gases into the air. Green energy evades the negative environmental impact; thus, it is vital. It uses eco-friendly substitutes to produce energy and does not generate unwanted gases in the environment. It comes from natural resources and is often clean and renewable. As a result, it generates a number of few greenhouse gases (Shah 2021).

There are many examples of green energy that are used today (TWI, n.d.):

- *Cooling and Heating in Buildings:* Green solutions used for buildings. These include solar water heaters, biomass-fuelled boilers, direct heat from geothermal, and cooling systems powered by renewable sources.
- *Heat for Industrial processes:* Renewable heat runs using biomass or renewable electricity. Hydrogen is a significant renewable energy provider for the chemical industries, iron, and steel.
- *Transport: Renewable:* Electricity and biofuels are growing in use for transportation across various industry sectors. Automotive is an obvious instance as electrification advances to substitute fossil fuel (TWI, n.d.).

There is enormous demand for energy in the world, and the future is moving towards gaining energy efficiently as well as securely. There is a need to utilise renewable sources of energy to solve this problem. The objective of green energy is to safeguard the climate system, improve its policies, and implement prevention. Traditional energy sources cannot meet the present and future energy demands. Green energy is an energy that does not damage the environment and is generated from many natural resources, for example, sunlight, water, and wind (TWI, n.d.).

Presently there is no high involvement of renewable energy in the world. Developing and developed nations both continue to use fossil fuels in the coming future. In developing countries, the situations are more problematic than in developed nations. Various developing countries have been directed to restructure their energy sectors (Salvarli and Salvarli, 2020).

3. Differences between Green Energy and Renewable Energy

There are some differences between green energy and renewable energy. The sources of green energy are primarily renewable energy technologies, for example, solar energy, biomass, geothermal energy, wind energy, and hydroelectric power. These technologies work in diverse ways, whether by taking power from the sun, using wind turbines or solar panels, or the flow of water to generate energy (TWI, n.d.).

Green energy is natural instead of fossil fuels like coal or natural gas, which can take millions of years to progress. Green energy does not produce pollution, which is due to fossil fuels. Therefore, all sources used the renewable energy are not green. For example, co2 produced due to organic material burns from the sustainable forest during power generation may not be green. Green energy sources also often avoid drilling or mining operations that can harm ecosystems (TWI, n.d.).

4. Green Energy, Clean Energy, and Renewable Energy

People often use green energy, clean energy (Figure 4), and renewable energy interchangeably. However, there is a difference between these terms (Table 1). For example, renewable energy may not be green or clean (such as some kinds of biomass energy).

Table 1. Green Energy/Clean Energy/Renewable Energy

Green Energy	Clean Energy	Renewable Energy
Green energy is the energy that comes from natural sources such as sun.	Clean energy is the energy that do not release pollutants into the environment.	Renewable energy is the energy that comes from sources that are continuously being replenished such as wind, solar energy or hydropower.

Renewable energy is the energy that comes from sources that are continuously being replenished. A source such as wind power is green, renewable, and clean since it occurs from an environmentally friendly, non-polluting source and is self-replenishing. Let us understand with another example: Can a hydroelectric dam that may distract waterways and impact the local environment be green?

Green energy is part of the future, offering a cleaner alternative to many current energy sources that are unsuitable for the environment but also lead to job creation, economically feasible, and development. Fossil fuels do not provide a sustainable solution to energy requirements. A variety of green solutions produce a sustainable future for energy provision without harming the world (TWI, n.d.).

Green energy appears to be part of the future world, offering a cleaner alternative to several of today's energy sources.

Figure 4. Clean Energy.
Source: https://cdn.pixabay.com/photo/2022/01/21/20/54/clean-energy-6955730_960_720.jpg

Fossil fuels require to become a thing of the past as they do not make a sustainable solution to energy requirements available. By setting a variety of green energy solutions, it is possible to create a sustainable future of energy without harming the world.

The sources of green energy are mostly renewable energy technologies, for example, solar energy, biomass, geothermal energy, wind energy, and hydroelectric power. These technologies work in diverse ways, whether by taking power from the sun, using wind turbines or solar panels, or the flow of water to generate energy (TWI, n.d.).

Green energy is natural instead of fossil fuels like coal or natural gas, which can take millions of years to progress. Green energy does not produce pollution, which is found with fossil fuels. Therefore, all sources used the renewable energy are not green. For example, co2 produced due to organic material burns from the sustainable forest during power generation may not be green. Green energy sources also often avoid drilling or mining operations that can harm ecosystems (TWI, n.d.).

5. Types of Green Energy

Various types of green energy are coming from multiple sources discussed in the next chapter. Some of these varieties are better suited to specific environments or regions, which is why there is so various renewable energy that filters into the energy grid (Davies, 2017).

6. Green Energy and Sustainable Development

Energy is the challenge of sustainability (Figure 5.) concerning social, economic, and environmental parameters. The installation costs, renewable resources, and policy structure will be principal factors; environmental impacts from energy production and usage are local, and the significant implications linked to the transport of pollutants in the environment can occur on continental, regional, and even transcontinental scales. Thus, various environmental, economic, and development are linked with the transition to sustainable energy systems and resources.

Figure 5. Sustainability.
Source: https://cdn.pixabay.com/photo/2018/04/06/12/49/sustainability-3295757_960_720.jpg

At present, sustainability development and electricity demand are rising rapidly. The aims of energy policy, directing energy mix, efficiency, and environmental standards should develop to provide many rehabilitations on unlicensed electricity generation and renewable energy resources. There are various main elements of the policies are: (Salvarli and Salvarli 2020).

- To confirm better free-market prices than the feed-in tariff

- To provide priority to renewable energy when associated with the grid.
- To provide extra encouraging sales tariff or domestically generate parts of renewable energy power plants.

The energy system can be an important reason for the environmental impact on developing and developed countries. A sustainable global energy system should provide to optimise and limit emissions. The global economy and technology must develop in harmony with sustainable development. As energy consumption is mainly from fossil fuels, global environmental matters are significant. Creating as well as developing countries are planning to use appropriate energy systems and enhance human, social, economic, and environmental situations for sustainable development (Salvarli and Salvarli 2020)

Salvarli also emphasised various challenges such as demographic, economic, social, and technological trends for the long-term sustainability of the world energy systems. Energetic actions should be taken in energy efficiency and diversity, supply reliability, public trust, market-based climate change response, reflective prices, regional integration of energy systems, technological innovation, and development to attain a sustainable energy system. Salvarli also discussed government policies and suggested that government policies should be planned for the replacement, transportation, production, distribution, and usage of energy. Due to the challenges and energy-related environmental problems, countries should implement and strengthen them effectively and efficiently.

According to Salvarli, sustainability aims to protect the climate system, improve its policies and execute related preventions. The use and current energy supply are vastly unsustainable depending on conventional fossil resources generated in unstable countries. To meet the present and future demands for improving situations such as economic, social, human and environmental technological changes will be necessary everywhere. Some issues such as investment, work, leadership, innovation, and organisation need to focus on.

Three critical factors determine the energy future: world politics, energy policy, technology, economic situations, market development, and technology. There is a need to focus on the environment, cultural heritage, and rich natural sources to approve the energy requirements of a country. Energy generation, transmission, distribution, and trade should be supported using standardised equipment and materials. Coal usage produces risks in the local environment, greenhouse gas emissions and pollution. The emission of carbon dioxide is high for coal. Diversification and utilising the country's resources are the key components that approve low-cost energy supply and sustainability. The subsequent investments in the industry should be developed for clean technologies. Economic and political factors will affect the quality of the clean environment. Offer to use domestic renewable energy resources diversity such as solar, hydro, biomass, and wind. Geothermal should generate more Electricity (Salvarli and Salvarli 2020)

The global energy supply mix will include gas, oil, coal and low-carbon sources by 2040. To handle pollution and limit Co2 emissions, coal usage should be controlled. Renewable energies are economically sustainable and eco-friendly when compared to fossil sources. The use of hydropower can confirm many profits for water supply and irrigation in agriculture, but it has consequences for the aquatic ecosystems. Geothermal power plant releases low emissions and is sustainable when linked to conventional fossil fuel plants. Environmental damages can occur if the pollutants are emitted from the power plant. Thus, cooled geothermal fluids are injected back into the earth, and the environmental risk is reduced. The impact of wind power related to the environmental impacts of fossil fuels is relatively minor. However, the operation and siting of wind turbines may consequence in adverse health effects on people who stay in the area of wind turbines. The usage of solar energy is rapidly improving all over the world. Though, many solar thermal and installed power arrangements are expected to be the same for focus solar power systems. Bio-energy is produced from biomass which is clean energy relative to the kind of technology and biomass used (Salvarli and Salvarli 2020).

7. Implementing Strategic Strategies

Renewable energy will likely account for 30% of energy building globally in 2050. Energy is the critical element in driving social and economic development. For implementing sustainable strategies, renewable energies are a significant source of energy as it is clean and support the goal of sustainable development (Salvarli and Salvarli 2020). Therefore, renewable energy development is improved by preparing legislation and policies with essential incentives. The strategic plan for renewable energies is necessary to raise competitiveness, care for the environment, and secure supply. Though knowing about the significance of green energy, fossil fuels are usually used, adversely impacting the environment and sustainability. Green energy sources are selected to substitute fossil fuels to develop the energy structure and improve the safety of the energy supply. In rural locations, green energy resources can resolve the energy consumption problem and associate with the agricultural production process that improves farmers' income (Salvarli and Salvarli 2020). For implementing strategic strategies, it is vital to focus on green energy production and consumption of green energy with an urbanisation system and focus on the local rural location to develop green energy and fulfil the requirement of local energy. This way, it is possible to deal with energy and environmental problems.

8. Obstacles to the Development of Renewable Energy

Developing countries face energy challenges and have some benefits in attempting to restructure their energy sectors and build efficient technologies. However, the circumstances of developing countries are different from developed countries. A significant part of the population can have difficulties attaining essential energy services due to resource constraints. Traditional technologies will likely remain cheaper than sustainable energy technologies (Salvarli and Salvarli 2020). There are many challenges to developing a renewable energy system, which creates obstacles to accomplishing the goal of a sustainable energy system.

Renewable energy sources should be provided to meet the challenges of main energy types and sustainable development of any country. To attain future challenges, major types of renewable energy sources with massive potential are solar, wind, hydro and biomass. For the sustainable energy supply, many requirements such as climate compatibility, sparing use of resources, low risks, social equity, and public acceptance should be satisfied. Barriers linked to societal and cultural patterns must be banned; therefore, desirable and sustainable alternatives and various incentives will be required to achieve more sustainable lifestyles. The current economic system remains an obstacle to alter due to the belief in unlimited natural resources and constant economic growth.

Conversely, the present construction sector is a rather conservative industry. It is known that sustainable designs, new building materials, and construction methods and practices are only evolving and being implemented slowly. Another obstacle is the high cost and long payback period for the energy efficiency of buildings. There are a lot of problems and barriers while developing renewable energy. Some technologies have been industrialised and commercialised to some extent and concerning the technologies. The following obstacles to the development of renewable energy are classified into three groups (Salvarli and Salvarli 2020):

1. *Cost Barriers:* Conventional energy sources have lower costs and prices than renewable energies. The leading cause of renewable energy's high production cost is low productivity technology and small scale. As the production cost of renewable energy is higher than that of fossil fuels with the same technology, there are barriers to the commercialisation and distribution relative to renewable energy.
2. *Market Share Barriers:* The current development of renewable energy comprises the cost barriers. However, a developed market can achieve system operation reliability and falling production costs.
3. *Policy Barriers:* Policy implementation and policy enactment are distinctive components of policy procedure. In the future, there is a need to develop renewable energy on an industrial scale. Thus, depending on the support for the market share of renewable energy and policies have to be increased.

Market share, policy and cost are the three main obstacles to renewable energy development. In the strategic plans of various countries, sustainable development concerning the parameters such as social, economic, and industrial is aided by their energy policies. New technologies related to renewable energies will also contribute to reducing environmental costs; therefore, the energy system will work economically and securely.

Industrial countries have 28% of the world's population and use 77% of energy generation. Predictably, today's world population will rise 1.26 times to reach 9.7 billion in 2050. 90% of the population growth belongs to the developing countries. The developed counties will adopt energy conservation policies by 2050. However, growing counties aim to construct electricity-producing facilities (Salvarli and Salvarli 2020).

About 75% of the energy demand and 67% of supply in 2016 will be seen by fossil fuels. Coal is significant as an essential energy resource globally, and its uses will be improved by 27% over 20 years. It estimates that the reserve of fossil fuels will come to finish one day. Thus, alternative and renewable energies will be vital energy resources shortly. There are two problematic situations with fossil fuel: first, it will end, and second, it negatively affects the environment. The environment is already polluted due to industrialisation and human development. The situation will be a cause to produce new jobs to advance new industries. Sustainable development comprises renewable energy, energy policy, grid technologies and renewable energy applications.

Economic, partly environmental, political, and human life data are based on the present energy systems. The basic parameter is to save energy and employ domestic energy sources. There will be a close relationship between the practice of energy and the environment in the future (Salvarli and Salvarli 2020). This is a pivotal time for renewable energy," said the IEA's executive director, Fatih Birol. In 2020, the UK hit a new amazing renewable energy milestone.

Industrial plant development should also enhance the economy, support ecology, and save energy. Energy investments related to environmental protection need considerable financial resources. The success of any innovative technology will be quantified by the focus on cost-effectiveness and eco-friendly. The rising energy demand requires to be met by clean power production. Clean and affordable energy will provide advancement toward sustainable development goals. The developing trends and new insights open significant business opportunities for energy leaders and organisations to permit new and innovative technologies and make better decisions. It is a new trend in the technology field that may be recognised and categorised as renewable energy, advanced materials, nanotechnology, manufacturing technologies, ecosystem, life-science green energy and information society technologies. These technologies also support strategic sectors with high market growth and social materials. The involvement of renewable energy is not high to get the chief energy and electricity supplies. An increase in the renewable energy industry, technology enhancement, and appropriate cost reductions are starting related to government policy precision investment and private sector inventiveness (Salvarli and Salvarli 2020).

9. Future of Green Energy

In the future, there will be a rise in demand for power. Thus, several renewable energy sources will continue to rise (Davies 2017). Furthermore, as the world population increases, so makes the demand for energy to power our homes, businesses, industries and communities. Therefore, innovation and expansion of renewable energy sources are important to maintaining a sustainable energy level and protecting our planet from climate change. Renewable energy sources currently comprise 26% of the world's electricity, but as per

International Energy Agency (IEA), its share is predicted to reach 30% by 2024 (Davies 2017).

9.1. Renewable Energy Facts

- Solar PV will account for 5% of world demand by 2020 and up to 9% by 2030.
- Energy can be met by 95% renewable energy by the year 2050.
- By the year 2050 price Waterhouse cooper forecasts that Africa could run on 100% renewable energy.
- The price of solar PV panels has declined 99% over the last four decades.
- According to a US study, renewable energy creates three times more jobs than fossil fuels.
- The world renewable market is now worth over $250 billion. Investment in renewable energy has exceeded fossil fuel investment (Davies 2017).

9.2. Green Energy in Domestic Sector

The advantage of renewable energy in a domestic setting are:

- *Cut Electricity Bills*: Once paid for, the costs of installing a renewable energy system public can become less dependent on the National Grid and energy bills can be diminished. For example, air source heat pumps can be a great alternative to a gas boiler.
- *Get Paid for the Electricity Produce by the Public*: The public produces electricity for its energy consumption and receives payment. For example, the UK government pays for the electricity produced by the public.
- *Sell Electricity back to the Grid:* If it is produced in excess, it can sell back to the national grid.
- *Drive Electric*: Usage of electric vehicles. EV tariffs are not only cost-effective but likewise renewable.
- *Reduce Carbon Footprint:* Green renewable sources do not release carbon dioxide or other harmful pollutants into the environment.

According to Energy saving trust's solar panel page, a typical solar PV system could save about 1.5-2 tonnes of carbon per year (Davies 2017).

9.3. Green Energy Products and Green Energy Premium

It is significant to comprehend that the energy we consume will mix green, renewable and conventional energy. It is due to all energy sources in the electric grid being mixed, whether they come in the power transmission grid. From here, electricity travels to businesses and homes via the handful of regional grids stretching across the United States and Canada. Therefore, by purchasing green energy, anyone is not only purchasing green power but also paying a small premium, including the costs of putting more renewable energy into the grid. Reducing the carbon footprint linked with energy consumption is best for those interested in going green at home but do not have space for solar panels. It is the best and most affordable method to increase large-scale renewable energy investment and provides more businesses and households access to green energy (Davies 2017).

If anyone green is a supplier of green energy like just energy, they will pay a very small premium to purchase green energy products. It is often similar to the cost of a week's coffee or movie ticket in most markets. This cost pays for green energy projects such as developing and maintaining solar fields so that they can move towards a greener and cleaner future combinedly. As a result, this movement gains more funding and traction, less of the energy that the consumer will be generated by conventional methods, which are unsustainable, contribute to air pollution, worsen global warming and harm the environment (Davies 2017).

Fossil fuels are still concerning in the largest share of energy consumption and retain their rising trend worldwide. In these circumstances, environmental pollution is somewhat inevitable, while renewable energy plants do not directly contribute. In the future, the key energy sources will be aimed at becoming new and renewable. While fossil fuels are unavoidably running out, renewable energy is more significant because they are effective in various areas such as generating jobs, continuous cost reductions, developing future industries and meeting energy and environmental targets.

The development and usage of renewable energy will enhance the energy environment, economy, construction, mechanical manufacturing, industry, and transportation and assist in developing new jobs. In addition, wind, solar and biomass energies can meet local energy demands and enhance environmental protection. The present situation is related to the energy demand inspires an enormous market for renewable energy. It is predicted that the share of renewable energy in meeting global energy demand will improve to 12.4% in 2023.

In the more extended period, if the investment in renewable technologies continues, renewable will have the possibility to make significant contributions to energy requirements. Additionally, various technologies comprise fuel cells and biofuels and participate in transport, heat and electricity markets. The share of fossil fuels in the total primary energy supply is predictable to comprise around 81% of the total sum in 2023; renewable energy will approximately account for 30% of the structure in the world.

By providing a balanced resource variation of countries for the primary energy resources, the share of renewable and domestic energy resources in the production system can be enhanced to the maximum extent. It is also aimed in the present strategy plans for various countries; targets should be attained in time to develop, support, and encourage new environment-friendly practices in production and services. The most significant market share and the most advanced renewable energy technologies belong to the leading developed counties such as Japan, the USA and Europe.

Use less and cleaner energy in buildings, power plants, industrial facilities and transport systems, and various energy-efficient enabling technologies. These technologies could slash costs by up to 80%, confirm energy savings by up to 30% and assist in slowing global warming in the future. Thus, the countries could stay cost-effective and make sustainable progress.

Conclusion

Green energy is produced from natural resources, for example, sunlight, wind, water or sunlight. Green energy aims to protect the environment and meet future energy requirements, which leads to improving situations such as economic, social, human and environmental changes in technologies. Conventional sources of energy increase pollution in the environment, and these sources are inadequate for fulfilling the need for energy. Therefore, green energy is a solution to current problems. People often use clean, green, and renewable energy interchangeably, though there is a difference between these terms. Green energy is the energy that comes from natural resources, whereas renewable energy is the energy that comes from sources that are continuously replenished.

There are many hurdles in the way of a green energy system, such as cost, market share, and policy. A sustainable global energy system focuses on limiting emissions and optimising energy efficiency. Consequently, the global economy and technology must progress in harmony with the sustainable development of the energy system. Actions should be taken in the arena of energy efficiency, public trust, market-based climate change response, regional integration of energy systems and technological innovation and development, usage of green energy sources and formulation of government policies etc., to achieve sustainable energy systems (Salvarli and Salvarli 2020). It will also lead to slowing down global warming and protecting the environment.

It is the best way to use affordable methods to enhance large-scale green energy investment for business and domestic usage (Davies et al., 2017).

It is vital to focus on green energy production and consumption of green energy. A rural location is different from an urban location as both have diverse situations to implement strategic strategies. It is necessary to balance both to achieve sustainable development. Rural areas require energy for agriculture, local business, domestic, transportation, etc. Providing green energy in rural areas makes it possible to fulfil the energy required to bring social and economic stability. If we look towards the urban location, it is also necessary to develop green buildings, use green energy for industries, and use other green technology for industries and business services. The ultimate aim is to deal with the energy and environmental problems.

Government policies should be planned for transportation, production, replacement, energy usage and distribution. Due to the energy-related environmental difficulties, countries should strengthen effectively and efficiently (Salvarli and Salvarli 2020).

Chapter 4

Sources of Green Energy

Abstract

This chapter explains various sources of green energy or alternative energy. The key sources are solar, wind, hydroelectric, and tidal energy. As the world population is rising, the demand for energy to power homes, businesses, and communities. Green energy is the expansion and innovation in the energy field, which is crucial to maintaining a sustainable energy level and protecting the planet from climate change. It is a solution to offer environmental benefits and includes power generated by solar, wind, geothermal, biogas, low-impact hydroelectric, and certain biomass sources. It is a significant energy source because it does not release toxic pollutants into the atmosphere.

Keywords: green energy, solar energy, wind energy, hydroelectric energy, tidal energy

1. Introduction

There are various sources of energy. The two sources of energy are traditional and non-traditional. Green energy is also an alternative energy source and a solution to sustainability in the power grid. Moving from conventional to renewable sources is gaining attention in the energy field. The term green energy is commonly confused with renewable energy, but both differ. Green energy offers the highest environmental benefits and comprises power produced by wind, solar, biogas, geothermal, low-impact hydroelectric, and certain biomass sources. However, renewable energy includes the same sources as green energy. This energy generally comprises products and technologies that can substantially impact global and local environments. When anyone buys green power, they are also supporting many renewable energy projects. It also assists investment in technologies that helps them grow ("What Is Green Energy? Renewable Energy Source," n.d.).

2. Sources of Green Energy

The main sources of green energy are solar energy, wind energy, hydroelectric power, and tidal energy. Wind and solar power can be produced in people's homes or industries (TWI, n.d.) The renewable energy source is the sustainable source that is endless, for example, the sun. The term 'alternative energy' also refers to renewable sources, meaning alternative sources can also be non-sustainable sources like coal. Nuclear-produced electricity is not renewable, but it is zero-carbon energy. Like renewable energy sources, it emits low levels or almost no CO_2. Nuclear energy has a stable source as it is not dependent on the weather. Green renewable sources of energy do not release harmful pollutants into the atmosphere. According to the energy-saving solar panel (Figure 6.), a typical P.V. system save 1.5-2 tonnes of carbon per year (Davies 2017).

Figure 6. Solar Panels.
Source: https://cdn.pixabay.com/photo/2015/09/17/12/04/solar-panels-943999_960_720.jpg

2.1. Common Forms of Green Energy

Popular Renewable energy sources are solar energy, hydro energy, solar energy, geothermal energy, wind energy, biomass energy, and tidal energy (Davies 2017). Common forms of green or renewable energy are:

1. *Solar Power:* This source is generated using photovoltaic cells that capture sunlight and transform it into electricity. Solar power is used for hot water, cooking, lighting, and heating buildings. It is affordable for domestic needs (TWI, n.d.). Sunlight is the natural source of

energy. It is our planet's plentiful and easily available energy resource. The solar energy reaching the earth's surface in a single hour is more than the planet's total energy needs for a year. The amount of energy differs according to the time of day, season, and geographical location. In the U.K., solar energy is increasingly gaining popularity for energy usage (Davies 2017).

    a. Solar energy directly comes from the sun, and it is a clean source of energy. Generally, stars produce an unbelievable amount of energy through nuclear fusion. Nuclear fusion is the process by which small atoms are fused by pressure and heat to produce heavier atoms with much energy. This energy reaches solar radiation, which collects and is converted into usable electricity. Solar panels are the general form of solar energy gathering and are full of things named photovoltaic cells. When the light from the sun hits photovoltaic cells, they generate an electrical current through the photoelectric effect. An inverter then passes the current into alternating current ("What Is Green Energy? Renewable Energy Source," n.d.).

2. *Wind Power*: Wind energy uses the power of the flow of air around the world that pushes turbines that produce electricity (TWI, n.d.). The wind is a plentiful source of clean energy. Wind farms are an increasingly familiar sight in the U.K., with wind power making an ever-increasing contribution to the National Grid. Turbines drive generators that feed electricity into the National Grid to harness electricity from wind energy. Although domestic or 'off-grid' generation systems are available, not every property is suitable for a domestic wind turbine. Find out more about wind energy on our wind power page (Davies 2017).

3. *Hydropower:* It is also called hydroelectric power. This green energy source uses the water flow in rivers, dams, streams, or elsewhere to generate energy. It can also work at home using the flow of water through pipes or can come from rainfall, evaporation, or the ocean's tides (TWI, n.d.). Hydropower is one of the renewable energy resources that are most commercially developed. Building a large reservoir can generate a controlled flow of water that will generate electricity and a turbine. It is more reliable than solar or wind power and permits electricity to be stored for use when there is an increase in demand. Therefore, Hydro can be more feasible in various situations as a commercial energy source (TWI, n.d.).

4. *Geothermal Energy*: This energy uses thermal energy. It has been stored under the earth's crust. This resource needs drilling to access and has been used for bathing in hot springs for several years. This resource can be used for steam to turn turbines and produce electricity. This energy is also questionable to the environment because it requires drilling. This resource depends on the location for use (TWI, n.d.). Geothermal energy can be used to heat homes or to produce electricity. Geothermal energy is of slight importance in the Uk compared to other countries like Iceland, where geothermal heat is freely available (Davies 2017).
5. *Biomass*: This renewable resource also must be monitored carefully to be labelled as a 'green energy' source. Biomass plants use sawdust, combustible organic agricultural waste, and wood waste to create energy. Burning these materials releases greenhouse gas which is lower than petroleum-based fuels. In addition, biomass involves burning organic materials to generate electricity. Therefore, it is more energy-efficient and clear. By converting agricultural, domestic, and industrial waste into liquid, gas, fuel, and solid, biomass produces power at a much low cost (TWI, n.d.).
6. *Biofuels*: These organic materials can be transformed into fuels such as biodiesel and ethanol. It was 2.7% of the world's fuel for transport in 2010. It is estimated to meet over 25% of global transportation fuel by 2050 (TWI, n.d.).
7. *Tidal Energy:* This is a form of hydro energy. It uses twice-daily tidal currents to drive turbine producers. Tidal flow, unlike some other sources, is not constant. However, it is highly predictable and thus compensates for the eras when the tide current is low (Davies 2017).

3. Limit Usage of Fossil Fuels

Fossil fuels are not sources of renewable energy because they are finite. They also release carbon dioxide into the air, contributing to global warming and climate change. In comparison to coal, wood is slightly better. Wood is a resource that comes from sustainably managed forests. Wood pellets and briquettes are made from by-products of the wood processing industry (Davies 2017).

4. Increase Usage of Green Energy Products

It is necessary to understand that the energy people will be a combination of renewable, traditional, and green irrespective of which product they purchase. All energy sources in the power grid are combined when they enter the power grid. Therefore, when buying green energy, people are not directly buying green power for their homes but instead paying a small premium covering the prices of more renewable energy into the power grid.

Reducing the carbon footprint linked with energy consumption is the best way for people who do not have space at home. In addition, it is a reasonable way to enhance large-scale renewable energy assets and provides more households and businesses access to green energy ("What Is Green Energy? Renewable Energy Source," n.d.).

The world population is increasing the demand for energy to power homes, communities, and businesses. Expansion and innovation of renewable energy sources are crucial to maintaining a sustainable energy level and protecting the planet from climate change. Renewable energy sources make up 26% of the world's electricity. The international energy agency is expected to reach 30% by 2024. In the future, renewable energy sources and demand for power will increase (Davies 2017).

Energy is the current as well as a future need of the world. It is the property of objects that can be transformed into various forms or can be moved to other objects but cannot be created or destroyed. It is a renewable energy source that is generated to reduce its negative environmental impact. Renewable energy is those sources that directly come from nature, for example, sun, wind, rain, and tides, and it is possible to produce it again and again. These sources are abundant and are the cleanest energy sources on earth. Many renewable energy technologies comprise hybrid and related technologies. These are used effectively (Kalyani, 2015).

- Storing energy produced through renewable energy
- Predicting renewable supply of energy
- Contributing inefficient delivery of energy produced using renewable
- Energy technologies to consumers.

5. Usage of Green Energies

- *Solar Energy*: Sun is a huge source of energy that provides energy to all living creatures on earth. It is a clean energy source that generates almost 10,000 times more energy than the earth can generate in the 21st century. Solar energy can be converted into valuable energy directly using many technologies classified under two fundamental categories (Kalyani, 2015).
  - Photovoltaic cells: Solar energy is directly transformed into electricity by using photovoltaic cells. It is a new technology as solar cells were only first developed in 1975. Solar cells use light energy from the sun to produce electricity through the photoelectric effect. By using photovoltaic modules, the sunlight is transformed into a direct current (D.C.). This direct current is then transformed into alternating current (A.C.) with an inverter and adjusted to meet the power features of the utility grid or the load (Kalyani, 2015).
  - Solar Thermal: Thermal denotes the usage of the heat energy from the sun. Initially, solar radiation can be absorbed in solar 'collectors' to provide solar space or water heating at comparatively low temperatures (Kalyani, 2015).
- *Hydro Energy:* Hydro energy is a power that comes from the water cycle, a continuous process of falling and fast running water to produce electricity. It is an already established form of renewable energy that provides a major source of electricity, approximately 19% of the world's electricity. The majority of hydroelectric power is generated on a large scale in the world. However, there is a further possibility for small-scale hydroelectric projects. Small scale projects have less capacity and have small dams, which make less impact on the environment. In addition, micro-scale schemes generate power in kilowatts and are used in individual houses and small villages (Kalyani, 2015).
- *Geothermal Energy:* Geothermal is made up of two words; Geo means 'the earth', and the word 'thermal' means 'the heat'. Thus, geothermal is the energy generated in the form of heat produced from the radioactive decay of materials inside the earth. The first geothermal power plant was constructed in Larderello. Magma is

produced from the radioactive decay of potassium and uranium below the earth's crust and generates lots of heat (Kalyani, 2015).

- *Geothermal Resources*: Four varieties of geothermal resources are pressure, hydrothermal, magma, rock, and hot dry. Presently only hydrothermal resources are used commercially, and other technologies are still underdeveloped. There are three basic elements of hydrothermal resources: a heat source (magma), An Aquifer (container or water), and an Impermeable cap rock that seals the aquifer. Aquifers are drilled to blow out the geothermal energy, and steam and hot water are extracted.
- *Geothermal Technology:* Geothermal energy can be used in electricity production or space heating, greenhouse heating, laundries, industrial processes, and water heating. Geothermal systems are mostly subdivided into two production facilities disposal and mechanical system. Firstly, the hot water and steam are delivered to the surface through boreholes and wells. A mechanical system comprises a heat exchanger, pipe, pump, and controls to transport the energy for various applications. Therefore, a disposal system receives and stores the cool fluid in injection wells or storage ponds.
    o Geothermal energy is used as the heat source or heat sink. Various technologies are available: dry steam, binary cycle system, flash steam, and hot, dry rock and resources used at high temperatures.
- *Dry Steam Technology:* The dry steam power plant is appropriate where the geothermal steam is not mixed with water. Production wells are drilled down to the aquifer, and the superheated, pressurized steam is brought to the surface at high speeds and passed by a steam turbine to produce electricity. Then cooling towers are used to exhaust out waste heat. The competence of dry steam plants is affected due to gases such as $H2S$ and $CO2$ being reduced to 30%. The commercially available range is from 35 to 120 MW.
- *Flash Steam Technology:* Flash steam technology is used where the hydrothermal resources are in a liquid form. The fluid is sprayed and pumped into a flash tank with less pressure than the fluid, which results in hydrothermal fluid vaporizing. After this, steam is passed to the turbine and a generator to generate electricity. The greatest of the geothermal fluid is not flashed, so

it is reinjected into the reservoir or can be used in direct heat applications. If the fluids are at a high temperature, they are passed to a second tank which is then flashed into steam to produce electricity again.
- Binary Cycle Power Plants: Binary cycle power plants are used where the resources are inadequately hot to produce steam and comprise several chemical impurities. Alternatively, we can use the fluid remaining in the flash steam power plants in a binary cycle power plant. In the binary cycle procedure, first, the fluid is passed with a heat exchanger. Next, the secondary fluid, for instance, isobutane or pentane, taking a lower boiling point than water, is evaporated and then passed to a turbine to produce electricity. Then the residual fluid is reinjected into the ground.

- *Wind Energy:* Wind turbine energy produced by wind flow is termed wind energy. It is a renewable source of energy that can be used as a substitute for fossil fuels. Moreover, wind energy is clean energy that does not produce pollution or release harmful greenhouse gases. That is why it is considered one of the sources of green energy.
  - The wind is a form of solar energy caused by the sun's heating of the atmosphere, the earth's rotation, and surface irregularities. Wind turbines are generally installed in large land farms. As a result, wind power capacity increased to 369.553 MW by Dec. 2014 and quickly increased the total wind energy generation reaching about 4% of total electricity usage.
  - Usually, all the vast wind turbines have the same structure and contain a horizontal axis wind turbine having an upwind rotor with three blades. In a wind farm, each turbine is interrelated with a communication network and medium voltage power collection system.
  - Currently, a combination of variable speed generators is used. A wind turbine is a device that changes kinetic energy from the wind into electrical power. The smaller turbines are used for applications such as charging the battery, traffic lights signs, etc. While huge turbines are used for the domestic power supply, the power which is of no use sell back to the utility supplier by the electrical grid. Arrays of bug turbines, generally called wind farms, are becoming an increasingly significant renewable energy source. Various countries started using these resources to reduce their dependence on fossil fuels. Normally turbines are

designed so that they not only work during the blowing of wind but also at low wind speed.
- *Vibration Energy:* Vibration generated by a vast crowd or by massive traffic on the road, vibrations of long bridges, tall buildings, railroads, ocean waves, and tall buildings can also be harvested efficiently.
  - The vibration energy can be transformed into electric energy. It can be stored, which can be used efficiently in giving power to various low-power electronics appliances, and the huge vibration energy harvesting gain 1 W to 100kW or more. This mechanical phenomenon is called vibration energy when the vibration happens about a point at equilibrium. Everything currently in the world vibrates at some frequency, some at low, which is noticeable, and some are very high so that any human eye cannot detect them. Vibration energy comes into focus in current years as we are in need to find out various sources of energy that are continuous and renewable so that we can use these sources instead fossil fuels.
  - *Forms of Vibration Energy*

There are two forms of vibration energy

1. *Free Vibration:* When a mechanical device is a setup with an early input, it is free to vibrate, called free vibration. Examples of this kind of vibration are hitting a tuning fork and letting it vibrate. In this kind, the mechanical system vibrates for one or more periods than its natural frequency and ends with zero frequency.
2. *Forced Vibration*: When time-changing disturbances are implemented to some mechanical system, such as load, displacement, or velocity, the vibrations happen. The disturbances applied can be steady-state, periodic, transient, or random input kind. In these, the periodic disturbances can be harmonic or non-harmonic type example, the shaking of the washing machine due to unbalancing. Vibration energy can be transformed into electricity by using a transducer. There are two transducers available for this. One is piezoelectric materials, and the other is an electromagnetic transducer. Transducers are the devices that transform various forms of energy into electrical energy, and therefore,

in the case of vibration, mechanical energy is transformed into electricity.
- *Biomass Energy*
  - Biomass is a source of energy derived from burning plant waste and animals. All industries comprising forestry, agriculture, colleges/universities, hotels, sports venues, resorts, and correctional facilities, generate waste that can be transformed into heat and electricity. The September 2017 report by the U.S. Energy Information Administration forecasts that the capacity to develop bioenergy will enhance in 2018. Biomass fuels only emit the same amount of carbon into the atmosphere as was absorbed by plants in the passage of their life cycle. Biomass is an energy source that utilizes waste and transforms it into energy usage (Mcfarland 2017).

5.1. Advantages of Green Energy

There are various benefits of green energy. The most important advantage is that it does not emit harmful gaseous. Various benefits of green energy presents in (Table 2).

5.2. Application of Green Energy

There are various uses of green energy. Applications of green energy are:

- Use green energy in calculators, satellites, road signs, etc., as solar energy.
- Passive space heating by solar energy.
- Warm up and retain food fresh by solar energy.
- Wind-powered water pumps.
- In farming, like in fish farms, in the form of geothermal energy.

Green Energy has a future in almost every field of the world, like industrial, agricultural, medical, domestic, etc.

Scientists already have found many forms of green energy, such as solar, wind, hydro, etc. Moreover, now, they are working on new energy forms like radiation and biomass to reduce the usage of non-renewable energy sources as they are already depleting.

Table 2. Advantages of Green Energy

Solar Energy	Wind Energy	Hydro Energy	Geothermal Energy	Biomass
Solar energy is a clean source of energy. It doesnot emit harmful gaseous. Better source for future generation. Various everyday items such as calculator or other low power house using devices powered by solar energy.	It is a clean energy. It does not pollute air. Wind turbines do not produce atmospheric emissions that cause greenhouse gases or acid rain. It is renewable source of energy need require less cost. Land around wind turbines can be used for other users example farming. Combination with solar energy they can be used to provide steady supply of electricity.	Hydro energy is clean source of energy. It doesnot create by product during transformation. Hydro electric power is a domestic source of energy. It is vast, reliable and affordable source of energy. Hydropower efforts produce a number of benefits such as flood control, water supply and irrigation. Hydroelectric power plants reservoirs gather rain water which can be used for irrigation. Hydroelectric installations bring electricity, commerce, highways to communities, industry thus developing the economy access to education and health.	It is affordable and cheaper source of energy used for bath, heating, homes, preparing food, office etc. It has outstanding potential for mitigation of global warming.	Biomass is widely available, since our society steadily produces waste. It reduces the dependency on fossil fuels. It is less costly than fossil fuels. Adds a revenue source to manufacturers.

Conclusion

Green energy offers the greatest environmental advantage and comprises power generated by solar, wind, geothermal, biogas, low-impact hydroelectric, and certain biomass sources. Though renewable energy comprises the same sources as green energy, Common forms of green energy are solar panels, which use photovoltaic cells. The wind is an abundant source of clean energy. It uses the power of the airflow to push wind turbines, which produce electricity. Another source, hydropower, is one of the renewable energy sources that are most commercially developed—for example, building a large reservoir or dam. Geothermal energy can heat homes or produce electricity as it is the best renewable energy source. Finally, the biomass comprises energy used by waste. This chapter discussed all types of green energy. The ultimate goal of the world is to limit the use of fossil fuels that pollute the environment and increase the usage of green products. Due to the increasing need for energy, the world is facing an energy crisis, which can be resolved by finding alternative energy sources. Ecological energy is the new hope for it.

Chapter 5

Design of Sustainable Energy Systems

Abstract

To make efforts toward sustainable energy for all, sustainable objective number seven emphasizes the importance of ensuring that everyone has access to reliable, modern, and affordable energy. Policies that encourage active participation from all segments of society must be developed, designed, and implemented. The energy crisis is a major concern for many nations. As a result, both developed and developing countries must act in this direction. Information and communication technologies have ushered in a new era of global interconnection. The primary areas of global action include energy distribution, modern appliances, supply efficiency, grid infrastructure, transportation, agricultural, industrial, and construction activities. The two pillars of creating and implementing energy policy are renewable energy and energy efficiency.

This chapter explains how to construct a sustainable energy system that provides long-term solutions and develops a sustainable system for everyone.

Keywords: sustainable energy, sustainable goals, global action, sustainable policy, green building

1. Introduction

The world is getting closer to target seven thanks to signs that energy is becoming more sustainable and broadly available. Access to electricity has begun to increase in poorer countries. Efficiency in the power sector is rising, and renewable energy is bringing considerable benefits (United Nations 2018).

United National Environment Programme Goal 7 targets (United Nations 2018) are the following:

- Ensure everyone can access affordable, dependable, and cutting-edge energy services by 2030.

- Renewable energy's share in the global energy mix will have expanded considerably by 2030.
- By 2030, the global pace of progress in energy efficiency will have doubled.
- Improve international cooperation to enhance access to clean energy research and technology, comprising renewable energy, cleaner fossil-fuel technology, and advanced technology, and enhance investment in energy infrastructure and technology by 2030.
- Expand infrastructure and upgrade technology to supply modern and sustainable energy services for all developing countries, especially the least developed ones.

It is vital to do proper coordination between the nations to attain sustainable goals. Green energy is the solution to achieve the targets. According to Nilsson, the investment costs of green energy pathways are huge but profitable for society, and most of them are already in motion. Progress is slow but must be enhanced at national and regional levels. Feed-in-tariffs, white and green certificates, technology standards, and removing fossil subsidies are significant first steps already underway. These contribute to scaling and nurturing new technological regimes and unsustainable and destabilizing old ones (Vezzoli et al., 2018).

Information and communication technologies play a significant role in speeding the progress toward green energy and sustainable development. Due to the advancement of information and communication technologies, the world is facing a strong evolution that is more interconnected than ever (Vezzoli et al., 2018).

Today meeting the requirements of current generations without compromising the ability of future generations to meet their needs is called sustainability (Wikipedia Contributors 2019). The global action plans highlight sectoral action areas such as distributed electricity solutions, modern appliances, transportation, construction, industrial processes, agriculture processes, supply efficiency, and grid infrastructure, which address both power generation and major areas of energy consumption (Vezzoli et al. 2018).

Based on the facts and figures, 13% of the global population is still scarcely available modern electricity. Three billion people depend on coal, wood, and animal waste for cooking and heating. Indoor air pollution due to the emission of combustible fuels from household energy caused 4.3 million deaths in 2012. With girls and women accounting for 6 out of every 10 of these, and the share of renewables increasing at a fast rate since 2012 (United Nations 2018). These facts and figures reveal that due to non-renewable energy, various environmental problems can be solved with the increasing usage of renewable energy. Renewable energy and energy efficiency are the two primaries in the sustainable energy hierarchy. Both of them are pillars of the sustainable energy development policy. In many countries, efficiency is seen as a security benefit as it may be used to reduce the level of energy imports from overseas or foreign countries (Prindle, Bill, 2007).

2. Sustainable Energy System

There is a need to bring sustainability for the sake of the environment and well-being on earth (Figure 7.).

Figure 7. Sustainable Growth.
Source: https://cdn.pixabay.com/photo/2021/10/31/07/49/growth-6756491_960_720.jpg

2.1. Meaning of Sustainability

Sustainability is called sustainability by meeting the requirements of current generations without compromising the ability of future generations to meet their needs (Wikipedia Contributors 2019).

2.2. Sustainability and United Nation

The UN General Assembly recently designated 2012 as the International Year of Sustainable Energy for All. The decade of sustainable energy for all has been designated as 2014–2024. Ban Ki-Moon, the Secretary-General of the United Nations, convened a high-level group on the subject. They produced a Global Action Agenda ahead of the United Nations conference on sustainable development. However, the lack of energy jeopardizes the achievement of the Millennium Development Goals. Therefore, it is necessary to move toward a clean energy economy to avoid the dangerous warming of the planet (Vezzoli et al., 2018).

2.3. Energy and Sustainable Development

Three inter-related objectives were identified to achieve sustainable energy for all initiatives by 2030.

- To safeguard universal access to modern energy services.
- Double rate of enhancement in energy efficiency.
- Double the share of renewable energy in the worldwide energy mix (Vezzoli et al. 2018).

Sustainable energy for all defines specific needs for diverse contexts. Low and middle-income governments must develop conditions that enable growth by launching a clear vision, policies, national targets, regulations, and incentives that connect energy to the overall development and strengthen national utilities. The SE4A initiative brings together more than 80 countries from low- and middle-income countries. While developing countries must prioritize renewable energy and efficiency, developed countries must prioritize renewable energy and efficiency. They are supporting all three objectives externally through international action. They focus on existing plans to enhance the deployment of local renewable energy and improve energy efficiency throughout the entire value chain from primary energy-using energy services (Vezzoli et al. 2018).

3. Global Action for Sustainable Energy

One buzzword that gets bandied around a lot is "sustainable development." However, one of the most useful definitions of sustainability comes from the United Nations World Commission on Environment and Development: "Sustainable development meets the requirements of the present without jeopardizing future generations' ability to satisfy their own needs."

Three primary sustainability challenges emerging in companies make this form of development profitable.

- The world's population is growing exponentially.
- There is a finite supply of non-renewable natural resources available.
- Upwards pricing pressure on non-renewable commodities will impact sustainable business practices.

Africa has had six of the top twelve fastest-growing economies since 2011. Due to this increased socio-economic growth, food, shelter, and energy demand have increased. The rapid growth of networked technologies and distributed systems in Africa is amazing (Vezzoli et al., 2018). According to one school of thought in Africa, if Africa cannot sustainably generate its energy, it will be unable to eliminate poverty, reduce inequalities, and improve the well-being of its people. Africa has vast water and sunshine resources, which can produce cleaner, accessible, sustainable energy and is cheaper.

Contrary to this, over 600 million African people live in darkness without electricity. This deficiency of electricity has reduced the continent's economic growth and affected health facilities, the quality of education, particularly in rural areas, and agricultural activities. The challenge is to provide an opportunity to think about efficient, clean, resilient, and low-carbon technologies and sustainable development to decrease overdependence on fossil fuels (Vezzoli et al., 2018).

3.1. Kenya

Through the energy legislation of 2006 and the energy policy of 2004, the Kenyan government has made significant progress in building an energy development strategy. It allows a wide range of stakeholders, particularly the commercial sector, to participate in delivering modern and clean energy services. It also occurs when the country's petroleum resources have been discovered and will thus play a key role in diversifying the energy mix and tackling energy poverty (Vezzoli et al., 2018).

The SE4 action agenda for Kenya shows an energy sector-wide long-term vision across the period 2015–2030. It plans how Kenya will achieve her SE4All goals of having 100% universal access to modern energy services and increasing to 80% the share of renewable energy in her energy mix by 2030 and biomass Access to modern energy comprises energy for cooking and electricity. Kenya has selected the baseline year for electricity access as 2012.

Skills Development for the green economy and the green economy vision 2013 of the Western Cape Government (WCG) skills development is a knowledge-driven project championed by the CHEC-WCG coordinating and cooperating group on climate change (Vezzoli et al. 2018).

3.2. South African

The future of South Africa's economy is endangered by unemployment and poverty, declining and degraded natural resources, and the impact of climate change. Solving these problems lies in transitioning to a green economy characterized by low carbon emissions, social inclusion, and efficient use of resources (Vezzoli et al., 2018).

4. Moving towards Sustainable Energy Policy

There are several policy imperatives for both developed and developing countries:

- Improve energy efficiency and set minimum efficiency goals.
- energy subsidies should be redirected and reformulated. Determine the most plentiful renewable energy resources and implement rules to promote their long-term growth.
- Seek cooperation from wealthy countries in transferring contemporary energy technologies while developing indigenous human and institutional capacity to support sustainable energy systems.

- This will boost the availability of efficient, clean, low-cost cookstoves.

Policies are challenging to put in place. Individual consumers, local communities, private firms, industry, non-governmental groups, intergovernmental agencies, and donor organizations must all play an active role. Developing nations must take the lead on the new energy path, while wealthier countries must be prepared to provide assistance and recognize that they have a big stake in the outcome. The ability to connect and use diverse forms of energy has revolutionized living conditions for billions of people since the beginning of the industrial period, allowing them to enjoy the comfort and accomplish productive work (The social benefits, eere.energ.gov, 2022).

Renewable energy and energy efficiency are two top priorities in the sustainable energy hierarchy. They are both sustainable energy policy components. Energy efficiency is regarded as a security benefit in various countries since it may lower the quantity of energy imported from other countries while also slowing the rate at which domestic resources are exhausted. In addition, energy efficiency is a cost-effective method of arranging economies that do not automatically result in increased energy consumption. In the mid-1970s, for example, California began instituting energy conservation measures such as building and appliance standards with stringent efficiency requirements. As a result, California's per capita energy use has remained steady over the years, although national consumption has more than doubled.

California has placed a "loading order" for new energy supplies that prioritizes energy efficiency, renewable electricity supply, and new fossil-fired power plants. States such as New York and Connecticut have established quasi-public Green Banks to assist corporate and residential property owners in lowering emissions and energy costs. According to Lovins of the Rocky Mountain Institute, industrial systems have numerous opportunities to save 70% to 90% of the energy and cost of fans, lighting, pump systems, and fans. 50% for electric motors and 60% for cooling, heating, appliances, and office equipment. Generally, 75% of the electricity consumed in the United States could be saved with efficiency procedures that cost less than the electricity itself. It holds for home settings as well. The US Department of Energy specified that there is potential for energy savings in the magnitude of 90 billion kWh by enhancing home energy efficiency.

According to the McKinsey Global Institute Report 2006, adequate, economically feasible opportunities for energy-productivity enhancements exist that could keep global energy demand growth at less than 1% per year, less than half of the 2.2 per cent average growth projected through 2020 under business-as-usual conditions. Energy productivity measures the quantity and quality of goods and services produced per unit of energy input and can be obtained by either reducing the amount of energy required to produce something or increasing the quantity or quality of goods and services produced from a similar amount of energy (Efficient Energy Use, Wikipedia, 2022)

There are numerous reasons why developing countries can succeed as leaders in sustainable energy. The first is to provide a billion people with essential energy services. For example, the amount of electricity needed to read at night, listen to radio broadcasts, and pump a small amount of drinking water comes to only 50 kWh per person yearly. Even when multiplied by the 1.6 billion people who currently lack access to electricity, this additional consumption would amount to a negligible portion of the world's energy demand (The social benefits, eere.energ.gov, 2022).

5. Designing Sustainable Energy System

5.1. BIM (Building Information Modelling)

Building information modelling (BIM) is a process that involves the creation and management of digital representations of the functional and physical properties of buildings. BIMs are files that may be traded, extracted, or networked to aid in the decision-making process for constructed assets. Businesses, government agencies, and individuals use BIM software to plan, design, build, run, and maintain various infrastructures such as gas, refuse, communication utilities, bridges, tunnels, trains, and water. In addition, new technologies are emerging to help realize the goal of green buildings, which lessen the negative impact on human health and the natural environment. It includes safeguarding tenant health and employee productivity, eliminating waste such as pollution and environmental degradation, and efficiently utilizing water, energy, and other resources (Wikipedia Contributors 2019).

5.2. Green Building Programs

Some green building plans do not target retrofitting existing homes, although others do, particularly through public energy efficiency refurbishment schemes. Green construction principles are simple to adopt (Wikipedia Contributors 2019).

5.3. Sustainable Design Buildings

In 2009, the US General Services Administration published a report that identified 12 sustainably constructed buildings that are less expensive to operate and have excellent energy performance. Furthermore, tenants were happier with the building than with commercial structures. These are referred to as "eco-friendly buildings." (Wikipedia Contributors 2019).

In the building sector, Peter O. Akadiri, Ezekiel A. Chinyio, and Paul O. Olomolaiye present a conceptual framework focused on adopting sustainability principles. The concept is founded on the sustainable triple bottom line principle, which includes cost-efficiency, resource conservation, and design for human adaptation. Each principle, which includes strategies and techniques to be executed during the life cycle of building projects, is defined, and a few examples are provided to help clarify the procedures. This framework will enable design teams to balance social, economic, and environmental concerns to increase building industry sustainability. Public relations professionals use information. The construction industry is a vital component of every economy and has a significant environmental impact. In addition, construction is a major consumer of energy, water, and material resources (Akadiri, 2012).

There must be a commitment to environmental performance and applicable plans and initiatives to make building activities sustainable. The sustainable building method is critical to significantly contributing to long-term development. Sustainability is a broad and complex notion that has become one of the most pressing challenges in the construction business. The concept of sustainability entails enhancing people's quality of life by allowing them to live in a healthy environment with improved economic, social, and environmental conditions (Akadiri, 2012).

5.4. Natural Building

Natural building is typically done on a smaller scale, concentrating on locally available natural resources. Other subjects connected to green and sustainable design are green architecture and sustainable design (Wikipedia Contributors 2019).

5.5. Green Building

Green building, also known as sustainable building, refers to designing and implementing environmentally friendly and resource-efficient procedures, from planning to design, maintenance, operation, repair, demolition, and construction, and necessitates strong collaboration between engineers, contractors, architects, and clients across all project phases (Wikipedia Contributors 2019).

Green construction approaches supplement and expand on the traditional considerations of building design, such as economy, comfort, and durability. The three components of sustainability that must be considered across the supply chain are profit, people, and the environment. LEED (Leadership in Energy and Environmental Design) is a rating system designed by the United States Green Building Council for green building construction, design, maintenance, and operation (Wikipedia Contributors 2019).

Environmental considerations should have been taken into account by industry public relations practitioners. In implementing construction projects, the advancement of sustainable building and construction techniques aims to balance environmental performance with economic and social performance. The link between sustainable development and building and construction should be obvious because building and construction are both economically significant and have significant environmental and social consequences.

With increased environmental protection awareness, this issue has received widespread attention from public relations practitioners worldwide. Using sustainable construction and building methods has encouraged economic progress in the building and construction industry. Implementing sustainable construction and building methods has been promoted to foster economic development in the building industry while reducing the environmental impact of construction and building. The three principles have emerged: resource efficiency, design for human adaption, and cost efficiency to attain industry sustainability (Akadiri, 2012).

Conclusion

Energy problems are a severe matter for developed as well as developing countries. It is a global issue focusing on developing new policies and methods to overcome energy problems. Sustainable energy development is a significant theme that is discussed widely all over the world. However, there is a lack of awareness regarding tools, techniques, and methods used to make sustainable energy systems. It is impossible to move in this direction without having the proper knowledge to implement sustainable policies successfully. The global actions mainly focus on modern appliances, transportation, grid infrastructure, agriculture, and supply efficiency. However, there is a lack of green energy utilization due to high implementation costs. This cost will be recovered in the future. Developing new methods, techniques, and tools to implement the green design in all sectors is also necessary. Adapting policies and feasible systems according to the nation and creating awareness to make a green system successful is vital.

Chapter 6

Methods and Tools of System Design

Abstract

A green energy system is not easy to implement. There are lots of challenges in this field. Due to a lack of human skills, knowledge, and other resources, working on green energy system tools and techniques is not easy. Thought it is the need of today's world, as we face environmental problems and scarcity of energy resources. Collaborative work efforts are required from all sectors worldwide to perform in this field. Various government and non-government organisations are working in this direction.

There are differences in the characteristics of developed and developing countries, so it is necessary to understand them and work accordingly to develop and implement the design. Therefore, there is a need for efforts from all sides of the world to resolve the issues. This chapter explains the matters that need to be considered while developing a system design and discusses the current scenario of green energy system design and methods and tools for system design.

Keywords: green energy system, system design, bottom-up-approach, hybrid energy tools, green building, agent-based modelling

1. Introduction

In terms of economic nature, the informal economy is a significant source of business activity in most developing nations, even though it is not included in the Gross Domestic Product (GDP). Energy models often employ GDP as one of the drivers of energy demand; thus, incorporating the informal economy into GDP will dramatically alter the parameter values in the energy models. In developing countries, income disparity is greater than in industrialised countries. However, most existing energy models have ignored it by relying solely on average income. This is a prevalent misconception (Van Ruijven cited in al Irsyad et al. 2017).

Solar electricity will be able to be installed and used in the majority of residences. However, solar energy (Figure 8) is not appropriate in every situation, and not every home is suited. For example, a largely south-facing roof is usually required to maximise the output of solar panels. If the roof faces south-west or west, want some advantage, but it might be less effective and will not save as much.

Figure 8. Solar system.
Source: https://cdn.pixabay.com/photo/2020/02/21/10/55/solar-4867218_960_720.jpg

It is not wrong to say that various companies and organisations are working toward a green energy system and its products. To meet clients' variegated and growing demands, the energy system company developed a state-of-the-art infrastructure facility at their premise wherein they have facilities for designing, manufacturing, quality testing, research, and storage of our range. These functions are carried out under the guidance of skilled professionals, including engineered quality controllers, research associates, warehouse personnel, and sales & marketing executives. They work closely with clients to complete all these processes promptly and effectively. With the help of industrious professionals, it can undertake the installation & commissioning of domestic wind mills, beautification & solarisation, solar electrification, home lighting, and allied works (Figure 9) ("Green Energy Systems, Indore - Manufacturer of Solar Product and Solar Security System" n.d.).

Figure 9. Wind Mills.
Source: https://cdn.pixabay.com/photo/2021/06/03/13/18/windmill-6307058__340.jpg

A proper green energy system must be developed and designed to integrate green energy properly. It is the result of a collaborative effort by government and non-government organisations. However, there is still a lack of awareness of green energy system design. For the sake of the environment, it is essential to think it over and thoroughly understand the system.

2. Green Energy System, Company

Since 2004, the company "Green Energy Systems" has been working in non-conventional energy, partnering with like-minded individuals and focusing on energy solutions. Their focus was on energy audits, delivering eco-friendly solutions, and boosting the efficiency of energy gadgets now in use. ("Green Energy Systems, Indore - Manufacturer of Solar Product and Solar Security System" n.d.)

Green Energy Systems manufactures, supplies, and trades a high-quality range of Biogas Plants, Solar Products, Solar Security Systems, Solar Lawn Lamps, Solar Flood Lights, Solar Cameras, Solar Induction Lamps, Non-Conventional Energy Systems, Domestic Wind Mills, and Solar Inverters. Our extensive product line includes the Solar Cooker System, Solar Water Heating

System, Solar Water Heater, and Solar Power Pack. We also meet the needs of shutter siren locks, door siren locks, car siren locks, and two-wheeler siren locks. In addition, we offer a high-quality assortment of solar lawn lamps, solar single crystal lawn lamps, solar flood lights, and other products to our valued customers ("Green Energy Systems, Indore - Manufacturer of Solar Product and Solar Security System" n.d.).

In 2004, a highly skilled team of experienced IT professionals, R & D team members, and engineers formed the "green energy systems" to serve the expertise of interested or uninterested people in using non-traditional energy. In Indore, MP, the company is registered as a maker of non-traditional energy goods (thermal, wind, biogas, and energy-efficient items). The company also works as a dealer, distributor, and supplier of well-known brands (USHA-Mumbai, PV Powertech-Pune, Zytech-Gujrat, Enertech-Pune, Mailhem, Libra Pump-Rau, etc.) in the non-traditional energy field in India. ("Green Energy Systems, Indore - Manufacturer of Solar Product and Solar Security System" n.d.)

The company's major goal is to develop and provide high-quality non-conventional energy products at a reasonable cost. In addition, the company intended to teach people about entrepreneurship and self-employment by providing production maintenance and operative training. In addition, the company collaborates with an NGO called "AASS" (Association for Awareness of Social Systems) to raise awareness of non-traditional energy products. www.solarhome.co.in is a platform that provides online awareness and product purchasing services 24 hours a day, seven days a week. This website is overseen by highly qualified individuals ("Green Energy Systems" 2015).

The "Green Energy Systems" are also certified partners of government entities such as Madhya Pradesh Electricity Board (MPEB), Ministry of New and Renewable Energy (MNRE), and many well-known brands in India and the international market to provide a complete non-conventional energy solution. We have constructed a state-of-the-art infrastructure facility at our site to suit our clients' diverse and expanding demands, including capabilities for designing, manufacturing, quality testing, research, and storage of our range. Skilled employees are engineers, quality controllers, research associates, warehouse personnel, and sales and marketing executives overseeing these operations.

There are numerous questions and issues concerning deploying a green energy system. Organisations and businesses are working in this direction and

Methods and Tools of System Design 81

responding to people's inquiries. The following are some of the questions and their answers:

- Why Should Install an Air Source Heat Pump?

    There are two key explanations and benefits: lowering carbon footprint (the amount of carbon emitted when you heat your home) and lowering your heating and fuel expenditure ("Home and Business Energy Systems | Green Energy Systems" 2020).

- What is the process of an Air-source Heat Pump?

    A fan draws air into the air source heat pump and sends it as a liquid to a compressor. The compressor increases the temperature in this liquid even more before it is extracted and utilised to heat the water in the heating system, which then heats the home via radiators ("Home and Business Energy Systems | Green Energy Systems" 2020). This form of an air-source heat pump is known as an "air to water" system and is the most popular. Another "air to water" system, and it is the most popular, is the Another option (or office) is an "air-to-air" system, which requires hot air to be cycled around your home ("Home and Business Energy Systems | Green Energy Systems" 2020).

- What exactly is an air source heat pump?

    An air-source heat pump collects heat from the air, which is then transferred to the home's heating system to heat it (or office). As a fuel, air replaces gas or oil and is free ("Home and Business Energy Systems | Green Energy Systems" 2020).
    Air Source Heat Pumps can capture heat from the air at temperatures as low as 15 degrees Celsius. Mitsubishi Ecodan heat pumps transform 1kW of electricity into 3.2kW of valuable heat for the home, maximising efficiency all year round ("Home and Business Energy Systems | Green Energy Systems" 2020).
    An air-source heat pump can provide heat in the radiators to heat the home and heat the hot water in the immersion tank. However, the air source heat pump will not perform efficiently, so the alternative solution could be to utilise solar PV panels to heat the hot water.

Green technologies can install both technologies together ("Home and Business Energy Systems | Green Energy Systems" 2020). Plant room installation and ground source heat pumps are other technologies.

- What is the operation of ground source heat pumps?

The ground beneath our feet retains heat at a relatively constant temperature, giving a continuous and consistent supply of free heat. This heat is extracted by GSHPs and transferred to radiators (or underfloor heating) ("Home and Business Energy Systems | Green Energy Systems" 2020). GSHPs operate by moving a fluid (similar to antifreeze) around a subterranean loop of pipes. These pipes are usually laid horizontally in a loop within the property, but they can also be installed vertically up to 100m if space is limited. The heat in the ground is transferred to the fluid, which flows through a heat exchanger, recovered, and sent to radiators by the heat pump (or underfloor heating). A ground source heat pump can be installed to provide heat for radiators to heat the home and hot water for the immersion tank. However, if a ground source heat pump is heating hot water, it will not function as efficiently as it could, so an alternative answer could be to utilise solar PV panels to heat the hot water. A ground source heat pump must be installed indoors instead of an air source heat pump. However, the ground source heat pump can be installed outside one of our bespoke plant rooms. It will also allow relocating hot water tanks and other appliances to the bespoke plant room, freeing up space in the home. We can design a plant room to house everything the heating system requires ("Home and Business Energy Systems | Green Energy Systems" 2020).

3. Energy System Planning

Energy planning is increasingly focused on using energy system modelling and analysis technologies to develop scenarios showing decision-makers the implications and possibilities. Municipalities contribute to the transition to renewable energy systems by defining local policies and goals, but they typically lack the tools and resources to complete the necessary complex analyses. Traditionally, energy system analysis methods have been designed with either the breadth of national energy systems or detailed project-specific analysis, leaving local planners in the dark. Johannsen explored the planning procedures of four municipalities using a qualitative case-based method. As a result, it has been discovered that future municipal planning tools must combine the need for systematic studies with concrete and implementable initiatives and analytical complexity with operational simplicity (Johannsen et al., 2021).

The transition from storable and dispatchable fossil energy sources to variable renewable energy sources and sector integration makes transitioning the energy system to renewable energy supply challenging. With this transition comes a greater need for long-term plans and higher demands for analytical complexity and temporal resolution in energy system modelling, complicating the decision-making and transition-planning process (Johannsen et al., 2021).

4. Renewable Energy Design

The role of renewable energy techniques in creating and developing a sustainable framework for green building is discussed in this chapter. The first point of view is associated with the earlier structure and low-encapsulated energy building materials for designing and developing a green building framework. The major viewpoint is to control energy protection in green construction by utilising renewable energy strategies. Sustainable construction or green construction can be used interchangeably with green building. Durable construction means using environmentally responsible and resource-efficient procedures in development to ensure sustainability throughout the lifetime of the building. This chapter also presents the combination of renewable, energy-based technology for green building construction and sustainability with the economics of renewable energy (Tomar 2021).

A green building is created, constructed, built, operated, re-used, or restored in an environmentally friendly and resource-efficient manner. It is sometimes referred to as a "sustainable" building. Generally, a sustainable system can be defined as a living system that endures because resources are not depleted faster than they can be restored organically. In financial terminology, a sustainable economic policy is one in which expenditures are less than or equal to income.

A sustainable social system, in common parlance, can be defined as one in which members are empowered to create a synergistic whole. It is becoming increasingly evident that individual decisions impact the global scale and vice versa. As a result of energy efficiency and energy conservation, green building has grown in importance worldwide (Tomar 2021).

5. Application of Tools in Decentralized Energy Planning

An increasing emphasis on decentralised energy planning, strategies, and objectives at the city and municipal levels may spur the development of the energy system as a whole. Therefore, it needs the necessary capabilities to conduct energy planning, including access to appropriate energy system modelling tools. Numerous models and methods are available for the simulation of energy systems and the creation of transition scenarios. However, as Machado showed, modellers have primarily been interested in undertaking regional and national studies.

As new methods are developed to deal with the rising complexity of the renewable energy transition, modelling municipal energy systems offers a potential for additional research. Krog and Sperling believe that energy system modelling and tool use are only a small part of the overall energy planning process, particularly in the municipal context. Thus, tools and models should be available so their use does not hinder overall energy planning.

It is a prevalent misconception (Johannsen et al., 2021).

6. Green Building Design

Two other imperatives must be met with constructing a natural green building: using renewable natural materials and collecting quantitative data from the landscape to inform the architecture. The site must be chosen based on the original information. Rather than drawing inspiration from nature, architectural forms should not employ artificial colour. Natural construction techniques must make use of the materials available in the environment. Natural materials are highly recommended since they replace synthetic products, the manufacture of which uses enormous amounts of energy. Raw materials are referred to as "natural materials," and they can be employed in either modern or traditional artisan methods. The construction standard refers to the term "passive building," which can be done using various building materials. It can be regarded as a green building structure that promises the interior environment is as comfortable in winter without the need for a traditional heating system as it is in summer and is one of the prerequisites for achieving minimal energy usage. Passive solar energy capture, building orientation, and high-quality triple-glazed windows are required to construct a passive building. For example, by considering a structure's orientation and changing its windows to capture passive solar heat, architects and promoters must keep in mind. That energy consumption should be less than what it takes to create their designs, improving the aspect of sustainable development and improving daylight penetration, which increases employee productivity without incurring additional construction costs (Tomar 2021).

7. Hybrid Energy Tools

Adopting a single analytical tool is insufficient when studying a complicated system such as an energy system. Hybrid tools are the best way to analyse the system. Bottom-up models provide extensive specifications of energy demand, as shown in Table 1, but the demand is usually exogenous, with no interactions with energy costs, income, or other factors. Top-down approaches, on the other hand, often have a highly aggregated energy sector (Li et al., 2015). They are bringing all the varied power plant technologies into a single energy grid (al Irsyad et al., 2017).

8. Life Cycle Thinking Methods

Consideration of the Life Cycle Life cycle thinking approach is a systemic tool for comprehensively comparing the properties of renewable energy systems from the cradle to the grave. For example, life cycle costing (LCC) evaluates all a technology's direct costs. In contrast, environmental life cycle analysis (LCA) serves a similar aim but concentrates on estimating environmental impacts rather than monetary values. Therefore, when examining the sustainability implications of technology consumption in society, a triple-dimensional study of sustainability accounts for social LCA (SLCA) with the other two dimensions (LCA and LCC). For example, Manik et al. (2013) conducted an SLCA analysis to evaluate Indonesia's social effects on the palm oil-based biodiesel business. Their investigation included 24 parameters weighted by interviewing workers, local community members, society, producers, transporters, and mill owners in the palm-oil biodiesel supply chain (al Irsyad et al., 2017).

9. Developed and Developing Countries

Despite growing renewable energy targets for developing countries, many appear to have limited renewable energy policy options. They cannot automatically embrace global renewable energy policies since their features differ from those of developed countries. Consequently, analysing the inherent characteristics of developing countries is necessary to formulate an effective and efficient energy policy, but, unfortunately, most energy analytical tools do not incorporate those characteristics. This problem has been recognised and debated in several published review studies. For example, Bhatia, Bhattacharyya and Timilsina (cited in Irsyad et al. 2017), and Hiremath Urban et al. (cited in Irsyad et al. 2017) do not recommend economic approach-based energy models because the economic assumptions used are different from conditions in developing countries. Instead, they advocate using engineering approach models and detailed descriptions of energy systems.

On the other hand, top-down models can still be employed, according to Meier, Pande, Shukla, Van Ruijven et al. (cited in Irsyad et al. 2017), if the model's assumptions are modified to reflect the features of the developing country. Nonetheless, we noticed that those research talks on decision support analysis, life cycle thinking and system thinking techniques were limited. Nevertheless, according to our findings, these novel analytical methods have been widely employed in developing nations to analyse renewable energy systems.

Combining the new tools with traditional energy modelling will improve the robustness of the results when modelling and analysing energy systems. In the case of underdeveloped countries, all analytical tools can be used for analysis as long as appropriate adaptations are made. The accounting-based bottom-up model, for example, is suitable for projecting dynamic transition in emerging nations, but top-down approaches are the finest instruments for fiscal policy analysis. In addition, examining rural household engagement in renewable energy production could be used.

In underdeveloped nations, common strategies for adjusting the energy model comprise disaggregating energy sector data in top-down approaches and disaggregating single household sectors into poor and affluent families.

However, only a limited amount of research can consider the informal economy, whose data is challenging to collect. As a result, developing country energy policy analysis is being challenged to quantify the informal economy, which has resulted in bias in GDP variables in energy models.

10. Agent Based Modelling

Developing countries should produce hybrid agent-based modelling (ABM) suggestions that combine four perspectives: engineering, economic, social, and environmental challenges. ABM is a good modelling framework capable of generating heterogeneous agents in the simulation. For example, in social analysis, the agents may be homes with varying incomes, access to electricity, and location. In contrast, in engineering, the agents could be power plants with varying costs, emissions, and capacity factors. In addition, the hybrid ABM suggests optimal policies that consider not only generation costs and environmental restrictions but also macroeconomic repercussions and social acceptance.

To explain the interconnections between the two different renewable energy markets, the suggested model should represent both on-grid and off-grid renewable energy systems at the same time.

Furthermore, we recommend focusing on the number of power subscribers for general energy modelling in emerging nations. This indicator is required to assess the effects of economic restructuring or rural electrification projects on a country's power demand. Unfortunately, most existing estimates of electricity demand do not consider this signal (Adom, Arisoy, Ozturk, Atalla Hunt, 2016 cited inIrsyad et al., 2017).

The Johannsen study offers researchers and model makers helpful advice for developing energy system modelling tools for municipal-level energy planning. Furthermore, the practical experiences and insights gained through energy planning inspire other municipalities seeking to place a greater emphasis on energy planning and strategy development. Three concepts should drive the future development of city-level energy system modelling tools and municipal planners.

- Help with the progressive expansion of internal modelling capacity.
- Allow for the use of local awareness and knowledge.
- Please offer specific and actionable results.

Municipalities generally have a positive attitude toward energy system modelling; they recognise the significance of having internal modelling capacity and believe creating and analysing energy system scenarios is critical in creating future renewable energy initiatives. Most, however, lack the internal capacity to do such energy system modelling. Municipal planners typically have some scientific or engineering experience but lack energy-specific skills, most likely due to the planners' diverse areas of responsibility. In this case, municipalities only have a few people working explicitly on energy planning, and in some cases, no dedicated energy planners are available.

As a result, energy planning duties are frequently carried out by planners with diverse areas of competence. Energy system modelling tools must be accessible to municipal planners, or they will be excluded from energy planning practice. A solution could be to have a tool at different complexity levels. A basic level requires fewer input parameters than an advanced level, which requires a more extensive range of inputs, enabling more complex analyses and the provision of more detailed outputs. In simulation models, it is more evident that model outputs are the product of input data, restriction criteria, and assessment criteria selection. A simulation model, as opposed to an optimisation-based method emphasising one best option, may thus better promote the development of local energy system knowledge and awareness of the range of possibilities available and stimulate more active use of energy system modelling. Municipal planners are local specialists in terms of the features of their local energy system and the aspirations and strategies guiding the municipality's overall growth. Municipalities appear to take a pragmatic approach to energy planning, emphasising tangible steps and initiatives such as energy-saving measures or concrete strategic targets like energy demand reductions or CO_2 emission reductions. Tools must be in place to support this approach and assist municipal planners in building scenarios with relevant and actionable measurements and modifications at the municipal level (Johannsen et al., 2021).

Furthermore, tool outputs and findings should be concrete and visible, giving the user an immediate grasp of how the modelled scenarios affect strategic energy aims. More concrete and transparent results would also help energy planners engage in talks and partnerships with colleagues from other fields and engage with the public to improve support for energy programs through a greater sense of ownership and involvement. The existing concept of what constitutes energy planning is a significant constraint on the applicability of the study's conclusions. The findings must be viewed from a holistic energy transition plan, balancing the demands and requirements of numerous energy (sub-)sectors while maintaining a holistic overview of the overall energy system. It is distinct from traditional utility company growth planning or building-level energy efficiency studies, both referred to as energy planning but fundamentally different and hence require unique tools. The tool design concepts outlined in this study should be considered by tool and model developers when designing future tools aimed at the municipal scale to bridge the present gap between planning practitioners and model developers. Future studies would be beneficial in documenting how municipalities proceed with applying such tools that have been expressly developed for the context of municipal planning (Johannsen et al., 2021).

11. Bottom-up Model and Top-Down Model

The first is the bottom-up approach, and the second is the top-down approach. The bottom-up approach is focused on decentralising the energy system, whereas the top-down is a centralised system. There are advantages and disadvantages to both models, which are as follows:

11.1. Bottom-up Model

The bottom-up model has the following advantages:

- First, technical change and learning are endogenous variables.
- The detailed specification of energy sector technologies
- The detailed specification of energy demand

Weakness of the Bottom-up Model are:

- Limited diffusion behaviour
- Neglected macroeconomic linkages
- Exogenous power demand
- Minor price adjustments can have an enormous impact on the whole electrical sector.
- In practice, the effect is usually slow.

- Homogeneous marketplaces, in which the same technology costs the same regardless of market location.

11.2. Top-Down Model

The strengths of a top-down model are:

- Economic arrangements that are theoretically consistent with it.
- Policy feedback or response from all economic sectors.

Weaknesses include:

- A lack of representation from the power industry.
- Technical change is viewed as an exogenous variable in the form of an autonomous energy efficiency improvement parameter.
- The cost of technological change is defined as the elasticity of substitution; however, the elasticity is rarely calculated. The simulation result is just an extrapolation of the past (al Irsyad et al., 2017).

Every approach has some advantages and disadvantages, but if we consider the future, a bottom-up approach is best, decentralised and easy to understand the overall energy system.

12. Decision Support Analysis Tool

Analysis of Decision Support Systems Decision support analysis, such as multi-criteria decision analysis (MCDA), is another tool used to analyse the complexity of the energy social system (Pohekar and Ramachandran, 2004). When examining energy concerns, MCDA typically solicits input from diverse stakeholders. The analytical hierarchy process (AHP), preference ranking organisation method for enrichment evaluation (PROMETHEE), and elimination and choice translating reality (ELECTRE) are the most frequently utilised MCDA families (Pohekar and Ramachandran, 2004). AHP divides a complex system into three main objectives: the main objective at the top, criteria at each level, and sub-criteria at each sub-level.

Underlying information and stakeholder judgements, items at each level are then weighted to generate the priorities of each decision alternative. Similarly, PROMETHEE is a simple ranking system that ranks alternative actions from best to worst by applying weights and preference functions (Pohekar and Ramachandran, 2004; Behzadian et al., 2010). ELECTRE, on the other hand, employs binary outranking relationships to detect and eliminate unsuitable alternatives (Pohekar and Ramachandran, 2004; Vahdani et al., 2013) (al Irsyad et al., 2017).

13. Solar Panels

Solar panels are a fantastic way to generate renewable energy for the home that is also less expensive than fossil fuels. When paired with a battery, it is possible to be largely self-sufficient. Solar electricity will be able to be installed and used in the majority of residences. However, solar energy is not appropriate for every situation, and not every home is suited. A largely south-facing roof is usually required to maximise the output of solar panels. If the roof faces south-west or west, it will still get some advantage, but it might be less effective and will not save as much. Solar panels are incredibly effective for generating clean energy for homes and are cheaper than fossil fuels. When combined with a battery, it can be largely self-sufficient. Most homes will be able to install and benefit from solar power. However, solar power is not right for every situation, and not every house is suitable. Maximising what panels can produce usually requires a predominantly south-facing roof. If the roof faces south-west or west, people still get some benefit, but it may be less effective and might not get the maximum savings ("Home and Business Energy Systems | Green Energy Systems" 2020)

People do not generally need planning permission for solar PV systems. Exceptions are that if the individual property has a flat roof, is listed, or is in a conservation area, it is advised to check with the local council. Solar PV systems are most efficient without a storage solution if individuals use the electricity they produce during the day. So if someone works all day, leaving home empty, an individual may not reap the full benefits of his or her solar panels. However, solar could be a solution to powering a garden room or home office away from the main house ("Home and Business Energy Systems | Green Energy Systems" 2020).

14. Battery Storage and ESS

A battery storage system, also known as an energy storage system (ESS), is a large rechargeable battery that can store electricity from various sources for your home. It will allow a person to store all the electricity generated by his or her solar panels rather than merely using it to heat water. During the day, solar panels can generate more electricity than is required to heat water. In general, we use more electricity in our homes when the sun goes down than when it comes up, so we do not take full advantage of solar panels. The approach is to save that electricity for later use ("Home and Business Energy Systems | Green Energy Systems" 2020).

Renewable energy planners in underdeveloped nations should use analytical tools developed in rich countries with caution. Traditional energy usage, economic and demographic shifts, severe income disparity, and informal economies are some of the characteristics of emerging countries that may violate the assumptions of widely used mainstream analytical techniques (al Irsyad et al., 2017). As we shift to a low-carbon and green economy, the widespread implementation of renewable energy technology has become a global problem. Around 176 countries have stated their renewable energy targets, which are motivated by various factors (REN21, 2017). Renewable energy technologies are viewed as one of the paths to decarbonising the energy industry and lowering reliance on fossil fuels (Dannenberg, Taylor, Ozcan cited in Irsyad et al. 2017). Renewable energy has traditionally been used as a source of energy in rural parts of underdeveloped countries.

Conclusion

Energy planning and modelling have been based on fossil fuels and centralised electricity supply. Decentralised energy planning and approaches and decentralised energy system models must be developed and implemented according to important characteristics. Therefore, a bottom-up approach best matches the demand and supply of energy. It is also necessary to develop innovative tools, methods, and designs to fulfil energy requirements at all levels. These tools, methods, and designs should be developed based on environmental, social, and economic concerns. All are vital for the sustainable development of every nation, whether they are developing or developed countries.

The characteristics of developed countries differ from those of developing countries, but both aims are the same: to achieve sustainable development. All countries need to move towards clean technologies, green energy, green systems, and green tools. These are eco-friendly and do not harm the environment. Therefore, it is concluded that energy system planning and implementation should be carefully performed to achieve long-term goals.

Chapter 7

Building a Sustainable Energy System

Abstract

In the twenty-first century, the world faces energy concerns and challenges. Climate change is a topic that is being debated on a global scale. The most challenging task is meeting the needs of billions of people who still lack access to energy services while also attempting to cut emissions and safeguard the environment. Building security and sustainable design are essential topics to explore. In addition, there are numerous societal and financial advantages to sustainable design and construction. This chapter discusses the sustainable design and building for developing and developing countries and explains how to build a sustainable energy system, sustainable building, the need for global trends, and energy efficiency.

Keywords: sustainable energy, energy design, green building, technology advancement, global trends

1. Introduction

Sustainable design and construction both protect the environment and enhance energy efficiency. It is the topic of global debate to address existing and future energy and environmental issues. The most significant issue in the twenty-first century is satisfying the requirements of billions of people who currently lack access to energy services while also participating in a global transition to clean or low-carbon energy usage. Emissions from developing countries are rapidly increasing, causing environmental challenges such as poor air quality and climate change, which impede global prosperity. The importance of sustainable design and building security cannot be overstated. The social benefits of sustainable design at the national, community or societal level include enhanced environmental quality, reduced health risks from pollutants associated with building utilisation, and neighbourhood rehabilitation. Cost estimates for climate change mitigation progressively discover that energy efficiency improvements deliver the most significant and least expensive pollution reductions while improving efficiency to fulfil energy demand.

These issues require immediate attention. In addition, reliable, socially acceptable, and inexpensive energy services are required for reducing extreme poverty and meeting other societal goals ("The Social Benefits of Sustainable Design," n.d.)

Because of advancements in information and communication technologies, the globe is undergoing a significant metamorphosis that has resulted in a more interconnected world than ever. People, images, ideas, and money are disseminated more frequently and quickly. In the modern era, the concept of sustainable development has been linked to the availability of sustainable energy. Because of the richness of both human and natural resources, there is reason to be optimistic about tremendous growth as the twentieth century progresses. For the first time in 2008, more people resided in cities than rural areas, influencing emerging regions such as Africa (Vezzoli, 2018).

The United Nations launched the Sustainable Energy for All (SE4All) program in 2011 to ensure universal access to modern energy services, doubling the global rate of improvement in energy efficiency and tripling the amount of renewable energy. The purpose is to attain sustainable goals by 2030 (Vezzoli, 2018). Sustainable energy will free up resources for profitable investments in health and education and strengthen the renewable energy industry that has the potential to generate employment.

2. Sustainable Design and Building Energy System

2.1 Sustainable Building

Building security and sustainable design are critical topics to consider. According to the findings of this study, the new sustainable building had a beneficial impact on the occupants' well-being, sense of social belonging, job satisfaction, and other aspects of work-life that influence an individual's job performance. Also, a small increase in organisational performance occurred after the move to the new building ("The Social Benefits of Sustainable Design," n.d.). The social benefits of sustainable design at the society, community or societal level include enhanced environmental quality, reduced health risks from pollutants associated with building utilisation, and neighbourhood rehabilitation.

Sustainable buildings in building projects are intended to offer a framework for enhancing the quality and comparability of methods and techniques for evaluating the environmental performance of buildings. It identifies and recognises matters to be considered when using methods and techniques to assess environmental performance for new or current building properties in the construction, design, operation, refurbishment, and deconstruction phases. It is envisioned to be used in conjunction with, complementary to current assessment systems such as LEED, BEES, BREAM, etc. (Akadiri, Chinyio, and Olomolaiye 2012)

The sustainability needs are, to a larger or lesser extent, inter-connected. The challenge for designers is to convey these various sustainability needs together in innovative methods. The new design system must identify the impacts of every design and choice on the cultural and natural resources of regional, local and global environments. These sustainability needs will be related throughout the various stages of the building life cycle, during its useful life, from its design to managing building waste in the demolition stage. This framework puts the groundwork for developing a decision care tool to improve the decision-making process in executing sustainability in building projects. The complete decision support tool will be explained in the developed model for usage in the U.K. building industry (Akadiri, Chinyio, and Olomolaiye 2012).

2.2. Rate of Technology Enhancement

Emissions from developing countries are increasing rapidly and are creating environmental problems such as poor air quality and climate change, which also hamper the prosperity of people around the globe. Throughout history, humanity's energy use has been marked by four broad trends. The first is increasing consumption and moving from conventional sources to commercial forms of energy; steady enhancement in energy technologies; a tendency towards fuel diversification and decarbonisation, particularly for electricity generation; and enhanced pollution control. These tendencies are encouraging, but the difficulty is that the rate of technological advancement has not kept up with the negative consequences of rapid growth in demand. There is a pressing need to accelerate the development of energy efficiency, and low-carbon energy would result in reduced pollution and boost developing countries' economic growth. Sustainable energy policies enhance the development of traditional renewable-energy industries, which will have the extra benefits of developing new economic opportunities. It reduces countries' exposure to volatile world energy markets and protects their resources ("The Social Benefits of Sustainable Design," n.d.).

2.3. Need for Global Sustainable Trends

Human societies have conventionally defined energy usage by four broad trends: ("The Social Benefits of Sustainable Design," n.d.).

- Enhancing overall consumption as societies gain wealth, transition, and industrialise from traditional energy sources such as charcoal, wood, and dung to commercial forms of energy, mostly fossil fuels.
- During the twentieth century, there was decarbonisation and diversification of fuels used specifically for electricity production.
- Reduction in conventional pollutants associated with energy usage.
- Consistent gains in the efficiency and power of energy-producing and energy-using technologies.

Another historical trend expected to be crucial for future sustainability is the shift in the carbon content of the fuels utilised as primary energy sources. The transition of wood and other conventional biomass fuels from a reliance on coal in the early industrial age to a more modern energy mix that includes significant amounts of oil, nuclear power, and natural gas has significantly reduced overall carbon intensity. A successful response to the threat of climate change will necessitate a significant acceleration of the historical trends toward fuel diversification and decarbonisation, which will require action on a global scale. Large energy services represent innovation and economic opportunity in several emerging countries. With increasing affluence and a better understanding of the critical environmental and human health impacts associated with most traditional pollutants, energy end-use technologies such as cooking, stoves, automobiles, as well as energy conservation technologies such as power plants, have become increasingly cleaner, at least in terms of visible, local, and immediate pollutants. However, using traditional fuels such as wood and dung for cooking is unproductive and releases high pollution levels ("The Social Benefits of Sustainable Design," n.d.).

2.4. Sustainable Energy Agenda

The concept of sustainable energy includes many facets that different parties might interpret differently. For the sake of this research, we have defined sustainable energy as three pillars: energy security, environmental sustainability, and quality of life, all of which contribute to achieving sustainable energy, but none of which, individually or collectively, totally explains sustainable energy. Initially, the pillars are linked to the 2030 Agenda for Sustainable Development (2030 agenda).

2.4.1. Long-Term Performance Goals
Long-term performance goals (LPG) are measurable, sustainable energy targets critical to world harmony. The LPG represents the planet's expected state at a given time. They cannot be explained in such a way that the purpose of one region (or period) contradicts the goal of another. Setting a specific, measurable objective and accomplishing that goal can be used to define success. The limits must be explained as results in the year 2050 for the sake of this project (or 2100). The international pact to limit global warming (Figure 10.) to 2 degrees Celsius is an example of a target. The transition to a sustainable energy system infers far-reaching variations in the socio-economic system and may result in economic inequality and marginalisation, leading to conflict and tension. If not tackled carefully. The move to a sustainable energy system is a long-term undertaking and must embrace all pillars of sustainable development, looking to leave no one behind and sustain social cohesion.

The energy revolution needs an exponential amount of resources driven by the rising demand for electric vehicles, storage, and batteries. Therefore, the United Nations, international financial institutions, and all regional governments must direct investment toward low-income countries to accelerate the energy transition in eastern subregions such as Central Asia, the Caucasus, East and Southeast Europe, and the Balkans. In various cases, the legacy infrastructure is appropriate for the region to improve the interplay between renewable energy and gas and further embrace decarbonised gases' potential. In addition, as the energy transition accelerates, the social and economic impact of the transition will need to be addressed for all. It will require rethinking social values and quality of life aspirations ("Pathways to Sustainable Energy Accelerating Energy Transition in the UNECE Region," n.d.)

Figure 10. Global Warming.
Source: https://cdn.pixabay.com/photo/2017/06/04/04/16/warming-2370285_960_720.jpg

A recent report by the McKinsey Global Institute identified that half of the world's growth in emissions could be prevented at a negative net cost using energy efficiency measures. In particular, the report identifies that a world investment of U.S. $170 billion per year in energy efficiency would yield advantage of U.S. $900 billion yearly by 2020, providing an average internal rate of return on investment of 17% per year (Bressand, Farrell, and Hass 2007).

2.4.2. Green Building

A "green" building is one that, by its design, construction, or operation, lowers or eliminates negative consequences while also having the potential to have good benefits for our climate and natural environment. Any building can be a green building, whether a home, school, office, hospital, community centre or any kind of structure. However, all green buildings are not the same. Countries and regions have different characteristics, such as different climatic conditions, unique cultures and traditions, different building kinds, and wide-ranging environmental, economic and social priorities (World Green Building Council 2016).

Green buildings protect valuable natural resources while also improving our quality of life. Various features can make a building 'green', comprising renewable energy, such as solar energy, pollution and waste reduction measures. Moreover, good indoor environmental air quality and consideration of the environment in design, construction and operation, a design that enables adaptation to a changing environment, use of non-toxic materials, and efficient use of energy, water and other resources. World GBC supports its member Green Building Council and its member companies in individual countries and regions to pursue green buildings best suited to their markets (World Green Building Council 2016).

The Building Environmental Assessment Method (BREEAM) is a certification method that validates a building's sustainability. The DGNB system rates indoor environments and districts, as well as the sustainability performance of buildings. The World Green Building Council is researching the impact of green buildings on the productivity and health of their users, and it is collaborating with the World Bank through EDGE to promote green buildings in emerging economies (Excellence in Design for Greater Efficiencies). The Global Sustainability Assessment System (GSAS), Green Star in Australia, and the Green Building Index (GBI) in Malaysia are also utilised in the Middle East (Wikipedia Contributors 2019).

2.4.3. U.S. General Service Administration

Sustainable design principles comprise optimising site potential, using environmentally preferable products, improving indoor environmental quality, minimising nonrenewable energy consumption, protecting and conserving water, and optimising operational and maintenance practices. Using a sustainable design philosophy encourages decisions at every stage of the design process that will limit negative impacts on the environment and the health of the inhabitants without sacrificing the bottom line. It is a comprehensive, all-encompassing approach that emphasises compromise and sacrifice. An integrated strategy has a favourable impact on all aspects of a building's life cycle, including design, construction, operation, and decommissioning ("Sustainable Design," n.d.).

The 2005 Energy Policy Act (EPAct) addressed energy production in the United States and includes rules for "developing new government buildings to achieve energy efficiency at least 30% more than ASHRAE 90.1 criteria, assuming life-cycle cost is effective." A Memorandum of understanding was signed in 2006 by 19 federal agencies, agreeing to "government leadership in the design and construction and operation of high-performance, sustainable buildings." This document resulted in the Guiding Principles for Sustainable Federal Buildings, which charge agencies with enhancing building performance while maximising asset life-cycle value. Additional environmental management targets were created by the Energy Independence and Security Act (EISA) of 2007.

New GSA buildings and substantial modifications must fulfil requirements such as decreasing fossil-fuel-generated energy use by 65% by 2015 and 100% by 2030. Federal agencies must manage their buildings, vehicles, and overall operations to improve energy and environmental efficiency, eliminate waste, and save money. It requires agencies to achieve and maintain cost-effective goals such as confirming new construction, major renovations in buildings, efficiency needs, sustainable design principles, reducing potable and non-potable water consumption, and complying with stormwater management requirements. The Council on Environmental Quality announced revised guiding principles for sustainable government buildings in 2016. GSA evaluates and measures sustainable design achievements using the U.S. Green Building Council's (USGBC) Leadership in Energy and Environmental Design (LEED®) green building certification standard. LEED® is a set of conditions and credits that must be met for a building to be LEED ® certified ("Sustainable Design," n.d.).

2.4.4. Business Sustainability Concepts

Some examples of company sustainability are green space, sustainable design and construction, crop rotation, water efficiency fixtures, renewable clean energy, water treatment, and waste to energy recycling. Implementing one or more of these concepts becomes a key component of numerous companies' sustainability strategies (Burton 2021).

2.4.4.1. Green Space

Having green areas in a city is about more than just giving a touch of nature to the cityscape. Because of their environmental sustainability benefits, such as urban advantage, reduced heat buildup, reduced soil erosion, water quality protection, and enhanced air quality, developing green spaces can be part of a bigger design strategy.

2.4.4.2. Crop Rotation

It is the technique of planting various plants on the same plot of ground throughout several seasons. This method is utilised because the soil becomes depleted of specific nutrients when the same crop is grown in the same spot for several years. By rotating crops, a crop that depletes a certain nutrient from the soil can be followed the next season by a plant that returns that same nutrient to the soil. Some environmental benefits of crop rotation include nitrogen management, improved soil structure, reduced soil erosion, improved pest and disease control, reduced water pollution, and reduced greenhouse gas emissions.

2.4.4.3. Sustainable Design and Construction

Building depletes resources, generates waste, emits potentially hazardous emissions, and fundamentally alters the function of land, particularly its ability to retain and absorb water. Sustainable design and construction approaches aim to reduce or compensate for the negative impact. Aside from green spaces, examples include reducing nonrenewable energy usage, using environmentally preferable items, adapting existing structures, and conserving water. In addition, there are various advantages to sustainable building that contribute to the creation of a more sustainable future. They are as follows:

Ecosystem protection, better air and water quality, water conservation, emissions reduction, conservation and restoration of natural resources, temperature management, waste reduction, and waste stream reduction are all environmental benefits.

Economic benefits include optimising the building's life cycle, lowering operating costs, enhancing occupant attendance and productivity, raising property value, and contributing to expanding the green market.

2.4.4.4. Renewable Clean Energy

The benefits of clean, renewable energy are improved public health, economic benefits and inexhaustible energy, stable energy prices and reliability and resilience. Moreover, renewable clean energy is undoubtedly the most visible illustration of sustainability. Examples are:

- Solar Energy: When the electromagnetic radiation from the sun is caught, it generates electricity and heat (Figure 11).

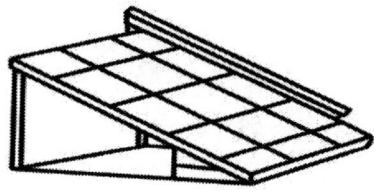

Figure 11. Solar Panel.Source:
https://cdn.pixabay.com/photo/2016/04/25/21/24/solar-panel-1353236_1280.png

- *Wind Energy:* Wind turbines transform the kinetic energy in the wind into mechanical power.
- *Geothermal Energy:* Heat escaping from within the earth can be used to generate electricity in geothermal power plants. These stations are especially appealing in seismically active places such as California and Iceland.
- *Waste to Energy Recycling:* As part of our sustainability strategy, we invested in waste-to-energy, which allowed us to save important fossil fuels. Moreover, it produces energy (electricity and steam) by burning non-recyclable hazardous waste solids (rags, organic debris, PPE, and absorbents) destined for landfills. Waste-to-energy will not solve all of the world's waste or energy problems, but it is an important first step. Terry's approach to waste-to-energy is an excellent example of corporate sustainability. At Terry, all waste is recycled, and nothing is disposed of in a landfill. Instead, the liquids are filtered and mixed before being routed to a solvent recovery. Solid waste is thermally treated at 1500°F to generate steam. That steam is used as an energy source to power those stills, reducing our dependence on valuable natural resources such as propane and water.
- *Water Treatment:* Water is a natural resource that is the free raw material used inefficiently by many sustainable enterprises. As the world's population expands, so will the demand for our water supply, and upward pricing pressure will surely drive up prices.

  This water treatment facility enables us to accept water-based acids, bases, oily water, and other water-based industrial wastes and transform them into clean water to meet our industrial water needs at our location. By processing this type of trash in our operations, we keep these waste streams out of landfills and eliminate the need for us to buy clean water.

Conclusion

The world is facing various energy-related challenges; thus, it is vital to focus on sustainable design and building. For this, it is essential to understand global trends and systems to make energy sustainable. Designing and building energy is not the job of one country alone. Thus it is essential to integrate all the policies and systems to resolve the global energy problems. Green building is the practice adopted by various countries, but all differ due to variations in social and economic factors. However, all are moving towards the same objective: the nation's sustainable development. It is very difficult to cope with current problems, develop policies, and prepare for the future. On one side, we struggle to fulfil the need for energy; on the other, we face climate change issues. However, if all sectors work together, nothing is impossible. It is building sustainability concepts discussed in this chapter which are green space, crop rotation, sustainable design and construction, clean energy, water treatment, and waste to energy recycling which require to adopt by all countries.

Chapter 8

Efficient Utilisation of Energy Systems

Abstract

Energy efficiency and renewable energy are two factors that contribute to a sustainable energy system. Therefore, energy conservation and energy efficiency are intertwined and must be linked seriously to use the energy system efficiently. It is possible to do this by applying more efficient manufacturing procedures and technology, as well as acknowledged procedures and techniques for reducing energy waste.

There is the moment to employ current tactics and technology for the benefit of both people and the environment. Modern appliances, architectural architecture, and LED lighting are some of the most recent initiatives to conserve and utilise energy. However, some challenges to this subject must be addressed. There is a need for an integrated energy system that focuses on minimizing the energy required while reducing the impact of air pollution. This chapter elaborated on the efficient utilisation of energy systems, energy conservation, energy efficiency, and ways to enhance energy efficiency.

Keywords: efficient energy system, energy efficiency, energy conservation, energy utilisation

1. Introduction

Energy efficiency is known as the efficient use of energy. It aims to diminish the energy required to provide products and services and reduce the impact of air pollution. For example, insulating a building allows it to use less cooling and heating energy to achieve and maintain thermal comfort. In addition, installing bright lighting, light-emitting diode bulbs, or natural skylight windows reduces the energy required to attain the same illumination level compared to traditional radiant light bulbs.

Energy efficiency improvement is usually achieved by adopting a more efficient production process and technology or applying accepted techniques to decrease energy losses. Several motivational factors take a vital role in enhancing energy efficiency. Decreasing energy usage reduces costs and may save consumers if the energy savings offset any additional cost of implementing an energy-efficient technology. Reducing energy use also solves the problem of reducing greenhouse gas emissions. ("Efficient Energy Use" 2022).

Energy efficiency cleans our air and saves energy and cost. Therefore, it should be urgent for every utility, individual, and government, stated John Nielsen, Clean Energy Program Director.

It is essential to combat climate change with the aid of energy efficiency. Energy efficiency is one of the simple and most effective ways to struggle with climate change. It saves consumers and businesses money and cleans the air (Gwen 2015). However, energy is wasted through inefficient technology and heat loss, costing businesses and families money and producing enhanced carbon emissions yearly.

International energy agency improved energy efficiency in buildings; industrial processes could reduce the world's energy need in 2050 by one-third and assist control emissions of greenhouse gases. Another solution is to eliminate government-led energy subsidies that encourage high energy consumption and inefficient energy usage in more than half of the nations in the world ("Efficient Energy Use" 2022). There are several questions in the mind of professional as well as researchers which makes it difficult for successful energy utilisation, such as;

- What makes impact an extremely efficient technology?
- How to motivate households and businesses to purchase and install new technology?
- How do driving performance and needless idling impact gas mileage?
- How many people will utilise public transportation if there is a cultural stigma against it?

Social, economic, and cultural differences create obstacles to the energy savings of high efficiency. Based on the research, 30% of the energy savings of high-efficiency technologies are lost. These issues significantly make our economy energy efficient and overall development (Dorsey 2019).

2. Energy Generation and Distribution

The strength of the renewable market lies in its variety and tangible benefits. For example, combined heat and power systems capture the 'waste' heat from power plants and use it to provide heating, hot water, and cooling to nearby buildings and services, enhancing the energy efficiency of power generation from approximately 33% to 80%. The smart grid is another system that will enhance the efficiency of electric generation, consumption, and distribution (Hole, Tom 2017). This infographic developed by the New Jersey Institute of technology's online master of science in electrical engineering degree program explores the diverse variety.

2.1. Community Design

Neighbourhoods designed with mixed usage developments and safe biking, accessible options for walking, and public transportation are significant in decreasing the requirement for personal vehicle travel.

2.2. Vehicles

Plug-in-hybrids fully electric vehicles are specifically fuel-efficient. More energy-efficient vehicles need less fuel to cover a given distance, producing fewer emissions and making them vital and less costly to function.

2.3. Freight

Freight can be moved more efficiently by enhancing the efficiency of rail and truck transportation and by moving long-distance freight transport from trucks to rail (Hole, Tom 2017).

2.4. Human Behavior

The four strategies above enhance energy efficiency mainly through technology and design. Though the way people use these technologies will strongly impact their effectiveness.

3. Energy Efficiency

The main motivation for energy efficiency is frequently simply saving money by reducing the cost of purchasing energy. In addition, from an energy policy fact, there has been a long tendency in a wider identification of energy efficiency as the first fuel meaning the capability to avoid or replace the consumption of solid fuels. The international energy agency measured energy efficiency from 1974-2010 and has eliminated more energy consumption in its member states than the consumption of any specific fuel, coal, including oil and natural gas. Energy efficiency brings other benefits with a reduction of energy consumption. Some estimates of the value of these benefits, called multiple benefits, ancillary benefits, co-benefits, or non-energy benefits, have placed their value even higher than that of the direct benefits of energy. These benefits of energy efficiency comprise things such s reduced air pollution, enhanced health, reduced impact of climate change, enhanced indoor situations, enhanced energy security, and decreases in the price risk for energy consumers. Methods for measuring the monetary value of these multiple benefits have been created, such as the choice experiment method for enhancement with a subjective element such as comfort or aesthetics and the Tuominen-Seppanen method for price risk drop. In the analysis, the economic benefit of energy efficiency investments can signify significantly higher than merely the value of the saved energy (cited in "Efficient Energy Use" 2022).

3.1. Meaning of Energy Efficiency

Energy efficiency means using less energy to work, similarly to eliminating energy waste. It has various benefits, i.e., decreasing greenhouse gas emissions, lowering costs, reducing demand for energy imports, and economy-wide. Renewable energy technologies assist in accomplishing these objectives. There are massive opportunities for efficiency enhancement in every sector of the economy, whether building, industry, transportation or energy generation. Moreover, this enhancing energy efficiency is the inexpensive and often the most instant way to decrease the use of fossil fuels (Dorsey 2019).

3.2. Ways to Attain Greater Energy Efficiency

There are various ways to achieve energy efficiency: (Gwen 2015)

- Using energy-efficient devices for space cooling and heating, lighting, and refrigerator; Enhancing system operations to improve various functions within a house, city, business, or other location and decrease energy usage.
- Designing our landscapes and buildings better.
- Moving towards energy-saving habits such as turning off the lights when there is no use.
- Adopting state, local or federal policies and programs can significantly increase energy efficiency. Program and policy options comprise:
- Building codes or other standards establish baseline energy efficiency for products that may be used or solely within a jurisdiction by building codes or other standards. For example, new homes require LED lights or minimum efficiency standards (Figure 12).

Figure 12. Energy Saving Lamp.
Source: https://cdn.pixabay.com/photo/2013/07/12/16/36/energy-saving-lamp-151234_960_720.png

- Presence of energy efficiency standards in municipal climate change or sustainability plans.
- With the aid of efficiency measures, various local governments have prepared climate change or sustainability plans, some of which comprise action items that decrease energy consumption.
- Programs to enhance energy efficiency. Some community-based organizations, utilities, or governments execute programs to educate consumers about energy efficiency and, in various cases, install or incentivize energy efficiency measures for residential or business consumers.
- Government, business, or organizations establish energy efficiency policies and their internal efficiency objectives (Gwen 2015).

3.3. Energy Efficiency and Conservation

Energy conservation and energy efficiency are interrelated, and energy efficiency usually pertains to the technical performance of energy alteration and consuming devices and building materials. The focus of energy conservation is to reduce the amount of energy end-use. For example, installing energy-efficient lights is an energy efficiency measure, while turning them off when not required, either manually or with timers or motion sensor switches, is an Energy Conservation measure ("Energy Efficiency and Conservation - US Energy Information Administration (EIA)" 2020)

Energy conservation and energy efficiency are both vital for effective energy utilisation. However, energy conservation is wider than energy efficiency in active efforts to reduce energy consumption, for example, through behaviour change and adopting good habits in addition to using energy more efficiently. Examples are heating a room less in winter, enabling energy-saving modes on computers and mobile, and air-drying clothes instead of using the dryer. It is a challenge requiring policy programs, behaviour change, technological development, etc. Various energy intermediary organizations, for example, governmental and non-governmental organizations on regional, local or national levels, are performing on often publicly funded projects or programs to meet this challenge. Psychologists have provided plans to change behaviour to reduce energy consumption while considering policy and technological considerations.

Energy conservation and energy efficiency measures assist in directly lowering energy costs for consumers and potentially reduce greenhouse gas emissions linked with energy use. When consumers reduce the electricity demand, they indirectly benefit and help reduce electricity production, distribution, and transmission costs. High demand for electricity results in a higher cost for power generation as well as transmission ("Energy Efficiency and Conservation - US Energy Information Administration (EIA)" 2020).

Examples of energy conservation and energy efficiency measures for consumers comprise of: ("Energy Efficiency and Conservation - US Energy Information Administration (EIA)" 2020)

- Purchasing energy-efficient products and vehicles. Installing energy management and control systems in industrial and commercial services. Contributing to utility energy conservation and energy efficiency programs that utilities provide their customers.
- Using programmable thermostats to control cooling systems and heating.
- Turning off electric appliances, lights, fans, and coolers when not in use.

4. Modern Appliances

Modern appliances such as freezers, ovens, dishwashers, cloth dryers, stoves, dryers, and washers (Figure 13) use less energy than old appliances. For example, currently, energy-efficient refrigerators use 40% less energy than conventional models. In Europe, if all domestic appliances changed, more than ten years would be protected with 20 billion kWh of electricity annually. Therefore, reducing CO_2 emissions by nearly 18 billion kg (20).

According to McKinsey and Company 2009, replacing old appliances is one of the efficient global measures to reduce energy usage by idle applications. Many countries identified energy-efficient appliances using energy input labelling. The impact of energy efficiency on top demand rest on when the appliances are used. For example, air conditioning utilises more energy during the afternoon when the temperature is high. Thus, an energy-efficient air conditioner impacts top demand more than off-peak demand. Conversely, an energy-efficient dishwasher uses more energy during the late evening when people prepare their dishes ("Efficient Energy Use" 2022).

Figure 13. Modern Appliances.
Source: https://cdn.pixabay.com/photo/2017/02/21/07/59/kitchen-2084994_960_720.jpg

5. Building Designs

Buildings are a significant field for energy efficiency improvement worldwide due to their role as an important energy consumer. However, energy usage in buildings is not straightforward as the indoor situations that can be attained with energy usage differ greatly. The measure that retains comfortable buildings, heating, lighting, ventilation, and cooling consumes energy. The level of energy efficiency in a building is measured by dividing energy consumed by the floor area of the building that is stated energy use intensity ("Efficient Energy Use" 2022).

Building designers are searching to improve building efficiency and include renewable energy technologies, creating zero-energy buildings. Variations in current buildings can also decrease energy costs and energy. These may comprise small steps such as selecting LED light bulbs and energy-efficient appliances or greater such as upgrading insulation and weatherization (Dorsey 2019).

Though the matter is more typical as building materials have embodied energy, on the other side, energy can be recovered from the materials when the building is dismantled by reusing or burning for energy. Furthermore, when the building is used, the indoor situations can differ, resulting in higher or lower quality indoor surroundings. In conclusion, overall efficiency is affected by the usage of the building. Therefore, a stable approach to energy efficiency in buildings should be more comprehensive than merely trying to reduce energy consumed. Factors such as indoor environment quality and space efficiency should be considered. Therefore, the measure used to enhance energy efficiency can take various forms. Passive measures integrally decrease the need to use energy, such as better insulation. Various functions enhance indoor situations and decrease energy usage, such as enhanced use of natural light. A building's location and environments play a main role in controlling its temperature and illumination. For example, landscaping, trees, and hills can block wind and shade. Tight building design comprising well-sealed doors, energy-efficient windows, and additional thermal insulation of walls, basement slabs, and foundations can decrease heat loss by 25 to 50%.

Dark roofs may become up to 39°C (70°F) warmer than the greatest reflective white surfaces. They convey some of this extra heat inside the building. US studies have represented that lightly coloured roofs use 40% less energy for cooling than buildings with darker roofs. White roof systems save more energy in sunny climates. Progressive electronic heating and cooling systems can produce reasonable energy consumption and enhance people's comfort in the building.

Appropriate placement of skylights and windows and using architectural structures that reflect light into a building can diminish the requirement for artificial lighting. Improved use of tasks and natural lighting has been represented by one study to enhance productivity in schools and offices. The choice of space heating and cooling technology in buildings can significantly impact energy efficiency and use. For example, substituting an older 50% efficient natural gas furnace with a new 95% efficient one will vividly reduce carbon emissions and winter gas bills. Ground source heat pumps can be even more energy-efficient and cost-effective. These systems use pumps and compressors to move refrigerant fluid around a thermodynamics cycle to pump heat in contradiction to its natural flow from hot to cold for the objective of moving heat into a building from the large thermal reservoir confined within the nearby ground. The heat pumps typically use four times less electrical energy to carry an equal quantity of heat than a direct electrical heater does the result. It can be reversed in the summer period and function to cool the air by transporting heat from the building to the ground is another benefit. Initial capital cost is the disadvantage of ground source heat pumps, but this is recovered within 5 to 10 years due to lower energy usage. The commercial sector is slowly approving smart meters to highpoint to staff and for internal monitoring of the building's energy usage in a dynamic presentable manner. Power quality analyzers can be used in the current building to use peaks, swells, assess usage, harmonic distortion, and interruptions to make it more energy-efficient. Repeatedly such meters communicate by using wireless sensor systems.

With the development of modern computer technology, many building performance simulation tools are accessible in the market. Green building XML is an evolving scheme, a subset of the building information modelling efforts emphasizing green building operation and design. It uses as input in several energy simulation engines.

6. Street Lighting

Around the world, cities light up millions of streets with the aid of 300 million lights. There are various ways to decrease energy usage in air transportation, from modifying planes and managing air traffic. As in cars, turbochargers are an effective way to decrease energy consumption, though instead of permitting the usage of a smaller-displacement engine, turbochargers in jet turbines function by compressing the thinner air at high altitudes. It permits the engine to function as if it were at sea-level pressures while taking benefit of the decreased drag on the aircraft at higher altitudes. Air traffic management systems are another way to raise the efficiency of the aircraft and the airline industry. New technology permits superior landing, takeoff, and collision avoidance automation within airports, from simple things like HVAC and lighting to more complex tasks such as scanning and security.

7. Efficient Utilisation of Energy

A vital part of the overall national energy addressees the efficient and smart end-use of electric energy.

Advanced data analytics and smart distribution systems will be investigated in this theme to meet enhancing customer demands, higher penetration of disseminated renewable resources, and the rise of retail electricity markets.

The main focus of this theme is to enhance the efficiency, controllability, and visibility of end-user energy consumption. In addition, technology development will have to reflect the customer's acceptance and participation. Thus, this theme will also address the social, behavioural, and psychological influences and human-machine interactions. Questions to address are: ("Power Utilisation and Energy Efficiency" n.d.)

- How to enhance customer satisfaction and participation in demand response and energy efficiency programs?
- How do we measure likely behavioural changes when customers familiarize a new technology?
- How to measure whether or not those fluctuations can sustain?

There is a need to do brainstorming on all of the above questions. Every problem has a solution, but the thing is to identify the problem. Once one recognizes the problems of the energy utilisation system, several solutions and suggestions come out. Some solutions are discussed below from the customer's point of view.

- Educate customers to increase their awareness of different programs
- Broad-mindedness of customers to short duration over-voltages and under-voltages in the presence of Distributed energy resources (DERs).
- Progress awarding or pricing schemes inspire customers to use more efficient appliances or participate in demand-side energy management agendas for moving energy and integrating renewable generation resources ("Power Utilisation and Energy Efficiency" n.d.).
- Develop demand response algorithms from the customers' perspective, such as centralized, distributed, and autonomous control algorithms.
  - Study pricing and rewarding schemes for improving consumer participation and acceptance of diverse energy efficiency and demand-side management programs.
  - Develop dispersed grid intelligence for providing a stage to implement the control, management, and coordination of the grid resources at diverse levels.
  - Develop prediction algorithms for commercial and residential loads and distributed production resources.
  - Develop advanced data analytic tools for the customers to manage building and home uses for monitoring device

malfunctions, providing suggestions for various consumption, security, safety, etc.
- Assess the quality, impact, and potential of the DR and efficiency programs, probable impacts of new loads (EV), acceptance of demand response programs, customer awareness, satisfaction, and acceptance ("Power Utilisation and Energy Efficiency" n.d.).

Customer contribution depends on letting customers comprehend the full advantages they will obtain when contributing to programs. Their satisfaction depends on cost-of-participants, rewards, and comfort. Therefore, the discussion centres around how to grow and develop a customer-oriented tool that can provide the following functions:

- Updating customers what the breakdown of energy use is to raise self-awareness and provide instant response to good behaviours.
- Forecasting what will be predictable if a DR or energy efficiency program is implemented.
- Advise desirable programs to customers and best practices to lower their energy bills and aid the environment while keeping and maintaining their comfort ("Power Utilisation and Energy Efficiency" n.d.).

8. Renewable Energy and Efficient Energy System

According to Gwen Farnsworth, Senior Energy policy advisor, Renewable energy is the way to reliable, clean power and affordable. It helps to power homes and cars with clean power, but to do so needs commitment and that strategic policies be implemented in the next five years.." (Gwen 2015). In 2015 renewable energy was reported for a tenth of the total US energy usage. Half of this was in the form of electricity. The most efficient forms of renewable energy are solar, wind, geothermal, hydroelectricity, and biomass (Hole, Tom 2017).

Biomass has the largest contribution with 50%, followed by wind power at 18% and hydroelectricity at 26%. Geothermal energy produces by harnessing the earth's natural heat. According to a recent report, the global industry is predicted to generate around 18.4 gigawatts by 2021 (Hole, Tom 2017). Wind energy makes use of airflow to move enormous wind turbines.

The mechanical action produces electric power. Rows of windmills are generally constructed along coastal zones where there is no obstacle to flow. This industry could make up to 35% of US electrical generation by 2050. Experts believe that solar energy could provide us with 25% of energy requirements. The approximate depends on jointly photovoltaic and solar thermal energy systems, which might not be far from reality given the continuing enhancement in solar technology and the stable reduction in the cost of the panels (Hole, Tom 2017).

Biomass denotes wood, waste, biofuels, and other kinds of organic materials which are burned to generate energy. The combustion procedure releases carbon emissions but is still well-thought-out renewable because the plants used can be regrown. Production will rise slower than the rest, from 4.2 quadrillions British Thermal Units (BTU) in 2013 to 5 quadrillion BTU in 2040, which presently accounts for 7% of the US total energy generation (Hole, Tom 2017). Hydroelectric plants utilise the power of moving water to produce electricity. The traditional method is to build dams to regulate the flow, which needs enormous investment, but function and maintenance costs are quite low.

Climate change due to carbon pollution from producing and burning fossil fuels will take a specifically hard toll on the desert west and mountain. The result of climate change are already noticeable; enhanced risk and duration of wildfires, drought, extinction of vulnerable wildlife species, and reduced snowpack. Traditional energy sources, for example, coal and natural gas, are enormous contributors to climate change. Electricity generation accounts for more than one-third of US global warming discharges, with the majority produced by coal-fired power plants. Non-renewable electricity production uses huge amounts of rare water and pollutes the air and water. In addition, reliance on traditional energy sources exposes consumers to price variations and harms our health and environment (Gwen 2015).

Renewable energy provides considerable advantages for climate, health, and economy. Most renewable energy resources permit no carbon pollution in the Mountain West abundant renewable resources that can be used to produce electricity like solar, wind, and geothermal. While geothermal energy systems permit some air pollutants, total air emissions are usually much lower than those of coal and natural gas-fired power plants (Gwen 2015).

Additionally, solar and wind need no water to function and therefore do not pollute water resources or rinse water supplies by competing with residential use, agriculture, or fish and wildlife. In contrast, fossil fuels can have a strong impact on water resources. For example, coal mining and natural gas drilling can pollute drinking water sources. Traditional hydropower is vital in the region, particularly in Arizona. However, the prospects for extra hydropower production capacity are inadequate due to environmental impacts, absence of potential location, and risk of insisted drought (Gwen 2015).

One feature that describes renewable energy sources is their fluctuations and intermittency. It covers both the expectable and unexpected fluctuations of their power outputs and uncertainty in the power accessibility. Special design thoughts should be considered to overcome the disadvantage of intermittency. These different designs enhance the overall cost of renewable energy systems. Adding a storage system or backup sources to the renewable sources is one of the measures used to guarantee the steadiness of power supply to the loads and thus enhance the reliability of the renewable energy systems. The function of a renewable energy system results in creating surplus energy, which is the energy produced by renewable energy systems but not consumed by the loads in standalone power systems. Effective utilisation of this additional energy can potentially reduce the cost of energy (COE) generation by these hybrid renewable energy systems (Ismail et al., 2015).

9. Measuring Renewable Energy Efficiency

The best renewable energy sources are calculating the costs of the fuel, the generation, and the adverse impact on the environment. It is followed in order by hydro, geothermal, nuclear, and solar. A formula was planned to compute the standard cost of electricity of the Levelized Cost of Electricity (LCOE) of the different methods discussed. The consequence depends on capital, projected utilisation, fuel, and operation and maintenance costs (Hole, Tom 2017).

Both plant owners and investors must reflect on the potential effects on the efficiency of other external issues. For example, there will continuously be a part of uncertainty when it comes to fuel prices and government policies. One administration may assist with tax credits and other stimuli for the industry. Another may not be as keen on seeing it take off (Hole, Tom 2017)

Aside from LCOE, the following formula used is referred the levelised evaded cost of electricity or Levelized Avoided Cost of Electricity (LACE) which measures the rate of the grid to produce electricity exiled by a new generation project. LACE seeks to address the gaps in LCOE by comparing technology efficiencies while accounting for regional variations (Hole, Tom 2017).

10. Vienna Climate Change Talks Report 2007

Vienna Climate Change Talks 2007 report under the support of the United Nations Framework Convention on Climate Change shows that energy efficiency can attain real emission reductions at a low cost. International standards ISO 17743 and ISO 17742 provide a documented methodology for measuring and reporting energy efficiency and savings for countries and cities. The energy intensity of a region or country, the ratio of energy usage to Gross Domestic Product, or some other measure of economic output, varies from its energy efficiency. Energy intensity is affected by economic structure, climate, trade, and the energy efficiency of vehicles, buildings, and industry ("Efficient Energy Use" 2022).

Climate change is not only an environmental problem but also a critical development issue. Its adverse impacts will disproportionally affect the poor, relying more on their immediate natural environment. In its long-term strategic framework 2008-2020 (Strategy, cited in adbheadhoncho 2010). ADB has identified energy as a core operational sector and achieving environmental sustainability as a strategic priority to pursue its mission to help its developing member countries reduce poverty and improve living conditions and quality of life. ADB has introduced new initiatives to augment its assistance to developing member countries in acquiring low-carbon technologies and implementing energy efficiency projects. ADB's country partnership strategy for energy operation in the People's Republic of China (PRC) has also emphasized clean and efficient technologies to promote the utilisation of renewable energy and help reduce greenhouse gas (GHGs) emissions (adbheadhoncho, 2010).

11. Western Resource Advocates

Western Resource Advocates were enhancing renewable energy and energy-efficient for clean air, cost savings, and curbing climate change. Western resources support works to enhance the usage of renewable energy, low carbon energy technologies, and energy efficiency. Western Resource Advocates protect the West's land, air, and water to ensure that vibrant communities exist in balance with nature. Western Resource Advocates Advances Energy Efficiency Policies, Programs, and Market Mechanisms. Western resources promote works to implement and develop policies, programs, and market mechanisms in the West to enhance the usage of energy efficiency and thus decrease carbon pollution from the power sector. Promoter before state public utility regulatory commissions and state legislatures. Promoters also compromise direct outreach to utilities, industry stakeholders, and community organizations. The goal is to enhance the usage of renewable energy, energy efficiency, and other low-carbon energy technologies so that by 2020, regional greenhouse gas emissions are 20% below 2005 levels and on the way to attaining an 80% reduction by 2050 (Gwen 2015). While these gains are vital, a more dramatic utilisation of clean energy technologies will be required if we address the problem of climate change. Western resource advocates explain a renewable energy vision by recognizing and extending best practices, encouraging technological innovation, analyzing the costs, and promoting more effective energy efficiency and distributed renewable generation (Gwen 2015).

Conclusion

Energy conservation and energy efficiency are interrelated, and both need to focus on efficiently using energy systems. Energy never destroys but can be transformed from one form to another. If it is not utilised correctly, then it is wasted. Proper utilisation can be done by using efficient production procedures and technology and ways to reduce energy wastage. Modern building design, LED lights, electric vehicles, and other modern appliances are steps toward saving energy and its efficient utilisation.

Moreover, regarding human behaviour and practices to save energy, Several issues need to consider for adopting the best energy system. First, there is a need for an integrated energy system that focuses on decreasing the quantity of energy required and reducing the impact of air pollution. Climate change is not only an environmental problem but also a critical development issue. Motivation is also required to adopt green energy to handle change management. Green energy helps move toward eco-friendly energy and save energy with new methods and techniques.

Chapter 9

Green Technology

Abstract

Green technology is new, but we are not aware of it. It focuses on the environment. It is to safeguard the earth and protect the environment. This technology provides solutions to the increasing demand for energy and environmental problems. It is time to discuss green technology to bring sustainable energy development. Green nano-technology employs green engineering and chemistry and is one of the modern green technologies. However, green technology implementation is problematic due to high costs and a lack of human capabilities. Establishing ways to deal with environmental and energy issues is vital to overcome this disadvantage.

This chapter introduces the notion of green technology and discusses basic principles of green engineering, green technology's application, and the advantages and disadvantages of green technology. Moreover, it explains the challenges of implementing green technology. Moreover, it also discusses the relevance of green technology as a sustainable solution and highlights several green technologies.

Keywords: green technology, nano-technology, green engineering, green energy

1. Introduction

Green technology is new, and various things are unknown. It is generally more expensive than the technology it seeks to replace because it does not account for the environmental costs inherent in many traditional manufacturing processes. In addition, the associated training costs and improvements make it even more expensive than other established technologies. The perceived profits concerning this technology are also related to other factors, for example, technology readiness, supporting infrastructure, human resource capability, and geographic elements (Iravani, Akbari, and Zohoori 2017).

Green technology (Greentech), environmental technology (envirotech), or clean technology (re) is the use of one or more environmental science, environmental monitoring, green chemistry, and electronic devices to monitor and preserve natural resources and the environment. Moreover, to mitigate the adverse effects of human involvement. Throughout history, animal power, human labour, water power, wind power, and firewood were the principal energy sources. People began to use fossil fuels in the 1800s. We are now reliant on fossil fuels. Though harmful energy policies, deforestation, excessive resource use, soil degradation, and climate change are issues that specialists must address to achieve sustainability on our planet can be supported by green technology (green machines 2020).

Green technology is the latest technology that everyone is talking about these days. However, we do not know what it means or how to apply it. It is a novel idea. Green technology, also known as eco-technology, is a type of sustainable technology encompassing practices, methods, and strategies for generating energy from non-toxic cleaning solutions. It is concerned with the environment and is suitable for the environment. It addresses energy efficiency, health and safety, renewable resources, recycling, and many others (green machines 2020). It is a term that describes a type of technology that is environmentally beneficial due to its manufacturing process or supply chain. The objective of green technology is to protect the environment and decrease environmental harm. Overuse of energy and pesticides, among other issues, has resulted in a greenhouse effect, global warming, and damaged habitats over the last two decades, promoting the development of green technologies. Environmental technology supports biodegradable materials, the creation of sustainable buildings, and recycling. It contributes significantly to carbon reduction, natural resource preservation, and global warming mitigation and allows people to consider environmental issues. As a result, in the future, this technology will deliver better solutions and be deployed in more efficient ways. It contributes to people's well-being and social success (Shafiei and Abadi 2017).

Green technology is becoming increasingly popular these days. Across the board, there is a growing interest in environmental issues. Restaurants that limit plastic, avoid plastic bags in the retail sector and avoid polythene are examples of transitioning to sustainable energy sources or purchasing carbon offsets. As a result, sustainability has increased throughout society (Long, 2019). The colour green connotes long-term viability. Green technology refers to environmentally friendly features. It focuses on sustainable innovation's long-term and short-term environmental implications (Long, 2019). This phrase is also used to describe renewable energy technologies such as wind turbines and photovoltaics. Environmental technologies develop on the foundation of sustainable development. The phrase "environmental technologies" also refers to a type of electronic gadget that can help with resource management in the long run ("10 Examples of Green Technology" 2019).

Green technology is applying technology and science to produce ecologically beneficial products. "Environmentally friendly" refers to any product or service that does not harm the environment. For example, it could relate to a service that has a tremendous environmental impact. There are several goals for green technology.

The primary goal is to conserve and preserve the environment. It may include an attempt to repair the damage caused by past or existing technologies. Bioenergy with carbon capture and storage is one example of a technology aimed at undoing previous damage Bioenergy with carbon capture and storage (BECCS). This technique converts crops into biofuels and captures the generated carbon dioxide. The majority of green technology is found in corporate and industrial settings. Examples include technologies that recycle trash, minimise pollution in water sources, and purify water and air. However, these are insufficient for industrial use and may also apply to household items (Abigail 2021).

The global problems of environmental degradation have forced society to rethink its way of development and evolve the concept of sustainable development. Indeed, new environmentally friendly technologies are critical to achieving long-term development. Various green initiatives are to maintain and improve the quality of the environment that might flourish in the new resource-efficient and sustainable-thinking society of the future. There is hope for international action in the application of science and technology to environmental concerns, born of the urgency of current environmental problems, the newly discovered recognition of mutual environmental interests, and the fundamental role of science and technology in general, and green technologies in particular, in assessing and responding to environmental threats. Shafiei and Abadi suggested using green technology, some of which are possible, as well as problems and alternatives in sustainable development. The administrators and professionals must use an appropriate model of a green building. To assess the success of green buildings, researchers identify difficulties related to developing green buildings and energy efficiency in industrialised countries. It can be revealed that green building is still in its infancy stage and such needs serious attention among players in developing green building and energy efficiency in developed countries. (Shafiei and Abadi 2017).

2. Basic Principles of Green Engineering

The term "technology" refers to a method, a collection of skills, techniques, methods, and processes used in the production of goods and services or in achieving goals such as scientific research. Technology is the understanding of connecting resources to create desirable products, satisfy, solve issues, and meet needs. Furthermore, it includes proficiency, technical methods, raw materials, and technical tools, or it can be rooted in machines, devices, computers, and factories that can be operated by people who do not have detailed knowledge of how such things work. The state of technology is the application of science, math, and art to benefit life. Monu Bhardwaj and colleagues, 2015) (Iravani, Akbari, and Zohoori 2017)

2.1. Criteria Fulfill by Green Technology

The criteria fulfilled by green technology are:

- It reduces greenhouse gas emissions to zero while ensuring safe usage and improving health and the environment.
- It conserves natural resources and energy.
- It slows down the deterioration of the ecosystem.
- It increases the use of renewables (Iravani, Akbari, and Zohoori 2017)

The term 'green technology' or 'eco-friendly technology' are broad concepts. However, they can contribute to the improvement of the environment. There are twelve principles of green engineering (Abigail 2021), which are:

- Reducing water
- Sustainability
- Developing eco-friendly products
- Efficient use of power increases the life-cycle of products
- Lowering the maintenance costs while enhancing the quality

- Promoting tools that decrease resource consumption in an indirect manner
- Limiting the usage of hazardous material

3. Examples of Green Technology

It is not simple to adopt green technology. Green technology varies from accessible devices such as programmable thermostats and LEDs to costly solar panels and turbines, with electronic vehicles somewhere between. There are tons of considerations about how renewable energy is being stored. Combating climate change needs a total disruption of society, involving the continued adoption of green technology (Long, 2019).

3.1. Energy Saving LED Lighting

An LED light bulb is an example of green technology that uses less energy than standard incandescent bulbs. Incandescent light bulbs were popular in the past, but they consume too much electricity wasted as heat. Then came LEDs and smart light bulbs, which revolutionised the lighting system. LEDs have a longer lifespan, require less energy and are made of safer materials. They also claim an 80% reduction in energy consumption over incandescent bulbs. It contributes to energy savings and lower electricity bills. These bulbs have a ten-thousand-hour lifespan. In addition, LED does not contain any dangerous compounds such as mercury. The next advantage of LED bulbs is that they concentrate light on a single spot, whereas incandescent and fluorescent bulbs transmit light in all directions. It can be operated via a home hub or a dedicated app when installed in a smart home. The most recent smart bulbs allow to control the time on and off and change colours.

LEDs have a longer lifespan, require less energy, and are made from safer materials. In addition, they claim an 80% reduction in incandescent energy consumption. It helps to preserve energy and reduce electricity expenses. These bulbs have a long lifespan of 10,000 hours. LED does not contain any hazardous compounds such as mercury. The next advantage of LED bulbs is that they concentrate light on a single point, whereas incandescent and fluorescent lights merely deliver light in all directions. When placed in a smart home, it can be managed via a home hub or a dedicated app. The most recent smart bulbs allow you to control the time on and off, change colours, and flicker on command (Abigail 2021)

3.2. Solar Panels

Solar panels are the most energy-efficient energy source (Long, 2019). Solar energy helps to cut both carbon emissions and electricity bills. Solar photovoltaic (PV) systems are sometimes known as solar panels. One of the alternative energy sources is a solar system. Fossil fuels power 62 per cent of the world's electricity generation, resulting in 50 million tons of CO_2 emissions yearly. As a result, greenhouse gas emissions are produced, damaging the environment. As a result, solar energy can reduce carbon emissions and electricity prices. Solar batteries can also store excess solar power for later use in the home. The batteries are linked to the solar panels. The electricity is subsequently sent to the batteries by the panels. An essential feature of these batteries is that they can transport electricity from the main grid, borrow electricity when needed, and return excess electricity (Abigail 2021).

Solar panels are the most famous type of green technology and may be small solar panels on an RV to run power when dry camping or solar panels on top of one's home to decrease energy bills. Solar panels are more common than solar concentrators. (Magazine 2004). Because solar panels are passive devices and more universal than solar concentrators, solar concentrators require significantly more money and space than solar panels. People have power when the sun goes down thanks to solar panels, even if they cannot charge the batteries throughout the day. Solar heating is a green technology that is commonly employed. For example, people save energy by using a solar cover to heat or preserve a pool without power. Greenhouses have traditionally relied on sun heating as well. The phrase "greenhouse effect" comes from (Magazine 2004).

3.3. Wind Energy

A wind farm is associated with wind power. It exemplifies green technology in action. On a small scale, windmills provide green technology in a home context (Long, 2019). Wind turbines are frequently named laterally, with solar panels as the gold standard of green technology. It can be installed in locations where solar panels are not operational due to low solar radiation levels. They may run day and night. The land under it can still be used for ranching and farming. The blades of wind turbines are rarely re-used. Wind turbines do not produce power when wind speeds are too low or too high (Magazine 2004).

3.4. Composting

Composting is the easiest and best green technology (Long, 2019). It means the breakdown of organic substances in the presence of aerobic organisms. It is a similar process that occurs whenever an organic substance is exposed to oxygen and moisture in gardens, forests, lawns etc. (Russo, 2005).

3.5. Vehicles

Advances in electronic vehicle technology recognise wireless charging capacities as the electric auto industry continues to evolve. Therefore, it is a better choice than a petroleum-powered car. As a result, electric vehicles (Figure 14) may be the future of the automotive industry (Long, 2019).

The most fundamental advantage of electric vehicles is that they can be powered by electricity, a readily available resource. It outperforms gasoline-powered vehicles in terms of efficiency. It is more efficient, requires less maintenance than internal combustion engines (ICEs), and can also accelerate. It provides a comfortable driving experience. Self-driving cars emit zero to minimal carbon emissions. Electric engines change energy electrochemically, and it does not generate toxic fumes like gasoline cars. It also runs on lithium-ion batteries, which are sometimes convertible for use with solar energy systems. However, the production of an electric vehicle generates some emissions. Another consideration is the source of electricity utilised to charge the vehicle. There would still be carbon emissions if the electricity needed to charge the car originated from fossil sources. When electricity is generated from renewable sources, it produces fewer carbon emissions (Abigail 2021).

Figure 14. Electric Vehicle.
Source: https://cdn.pixabay.com/photo/2017/07/27/13/07/electric-car-2545290_960_720.png

3.6. Programmable Thermostats

A programmable thermostat is a low-cost option for green technology. It can fix a schedule and adjust the temperature around comings and goings to save energy and money (Long, 2019). Programmable thermostats are another eco-friendly green technology and low-cost solution. They permitted them to devise a strategy for changing the temperature in residence. Most programmable thermostats allow setting a schedule for each day of the week. Set the thermostat, for example, to raise the temperature before leaving the house in the morning. Then, we can have it declined by the time we expect to return home in the afternoon. On weekends, the temperature can be maintained at a steady level.

Some thermostats can detect whether or not the home is occupied. It will then change the temperature mechanically based on pre-set levels. These features are frequently seen in smart thermostats. Then we do not need a smart thermostat at all. More choices allow us to set temperature adjustment schedules. People can save energy and money on utility costs by using programmable thermostats. Personal thermostats are also comfortable. They often allow for more precise temperature management. They also let remotely change and monitor temperatures using a dedicated app. They are also able to measure temperatures compared to manual thermostats more accurately. The energy-saving abilities of programmable thermostats translate into a reduction in greenhouse gas emissions. They also do not include harmful chemicals, for example, mercury. They are mostly electronic. That means we do not risk exposure to toxic chemicals that can harm our health (Abigail 2021).

3.7. Vertical Farming

Vertical farming is an environmentally friendly technology that involves the idea of producing in stacked vertical layers rather than horizontally. Therefore, it is vital to improving sustainability.

3.8. Wastewater Treatment

Technological advancements include microbial fuel cells, membrane filtration, nano-technology, biological treatments, and natural treatment systems such as wetlands. These methods are used to either prepare drinking water or dramatically reduce the presence of contaminants in what is released into rivers and the sea ("10 Examples of Green Technology" 2019).

3.9. Elimination of Industrial Emissions

The management and treatment of emissions of air pollutants in industries can significantly decrease the greenhouse effect. Carbon dioxide and methane are substances that damage the environment. For example, pharmaceutical, petrochemical, chemical, automotive, etc., industries must remove their omissions so as not to cause serious environmental damage. Technology is oriented to developing custom solutions for every company ("10 Examples of Green Technology" 2019).

3.10. Recycling and Waste Management

The increase in industrial waste and household waste has been disproportionate. Managing solid waste is the promise of individuals as well as companies. Technologies such as smart containers, automated optical scanning, and automated food waste tracking systems are outstanding technologies that can help sort mixed plastics by isolating them from others ("10 Examples of Green Technology" 2019).

3.11. Self-Sufficient Buildings

Self-Sufficient buildings can function by themselves without the requirement of an external contribution. To achieve greater productivity with the same surface of photovoltaic panels is to combine intelligent solar tracking systems, therefore obtaining the usage of radiation. ("10 Examples of Green Technology" 2019).

3.12. Waste-to-Energy Conversion

The production of energy from waste, also known as waste-to-energy, is a technology that produces energy from garbage. For each company's internal processes, waste treatment solutions generate energy from hot water, steam, or electricity ("10 Examples of Green Technology" 2019).

3.13. Production of Energy from the Waves

The first wave energy management plant was built in Agucadoura, Portugal, which is 8 kilometres away from the coast. The plant has a volume of 2.25 MW and can supply electricity to up to 1500 homes. The installation involves steel tubes floating on the ocean surface measuring 3.5 m in diameter and 150 m long, called a Pelamis. ("10 Examples of Green Technology" 2019).

3.14. Ecological Vehicles

Vehicles that do not release gas are called "ecological vehicles" because their use does not have a negative influence on the environment and involves reducing the presence of polluting gases in the atmosphere, primarily carbon dioxide (CO_2), nitrogen oxide (NOx_0), carbon monoxide (CO), unburned hydrocarbons (HC) and compounds of sulfur dioxide and lead ("10 Examples of Green Technology" 2019).

3.15. Harnessing Solar Energy

These are the systems that have been operated on and researched the most. Examples of solar energy alteration technologies are high vacuum tubes for hot water, photovoltaic collectors to generate electricity, polypropylene collection for hot water, and solar street lamps. These technologies focus on reducing dependence on energy from hydrocarbons and fossil fuels and enhancing greener solutions ("10 Examples of Green Technology" 2019).

Figure 15. Eco-friendly Technology.
Source: https://cdn.pixabay.com/photo/2022/01/28/07/48/tree-6973905_960_720.png

3.16. Vertical Farms and Gardens

Vertical gardens installed in buildings save energy and have numerous environmental benefits (Figure 15). Vertical gardens do not necessitate watering routines that wastewater. Because they are put along a wall, they reduce the extreme noise pollution that originates from the outside, and even that one can make. It also helps with the high temperatures caused by climate change, resulting in significant savings in heating, energy, and air conditioning. This method can save water while caring for the fertile soil ("10 Examples of Green Technology" 2019).

3.17. Green Boilers

Green boilers use as little fuel as possible or are powered by renewable energy. Although natural gas is a fossil fuel, it emits no harmful gases like carbon monoxide, particulates, nitrogen oxides, or sulfur. It produces less $CO2$ and more water vapour. Nowadays, modern technologies are available for both families and industries. Much of the work focuses on reducing pollution and waste ("10 Examples of Green Technology" 2019).

4. Application of Green Technology

Various sectors use green technology. Sectors using green technology are:

4.1. Energy Sector

The bulk of the world's energy is produced by burning fossil fuels. Green technology can provide more environmentally sustainable fuel sources than fossil fuels. As a by-product of their production, fossil fuels typically generate trash. Solar, wind, and hydroelectric dams can be utilised instead of fossil fuels since they are less detrimental to the environment and produce no hazardous by-products (Qamar and Mariya 2021).

4.2. Transportation Sector

Conventional fuel-powered vehicles are a major source of global GHG emissions. As a result, many firms are combining green technology into transportation infrastructure and vehicles, such as electric automobiles and compressed natural gas (CNG) buses (Qamar and Mariya 2021).

4.3. Waste Management Sector

Green Tech is also used in waste management to transport, store, and recycle waste (Qamar and Mariya, 2021).

4.4. Water Filtration

Green technology is being used to purify water all over the world. For example, it could be used to cleanse contaminated water or remove salt from seawater where water suppliers are scarce to boost the accessibility of safe drinking water (Qamar and Mariya, 2021).

4.5. Air Purification

Green tech is also being used to clean the polluted air by decreasing carbon emissions and gases released by the industrial sectors. (Qamar and Mariya 2021).

Although the green technology business is still in its infancy, investment capital is already pouring in. While green technology has become increasingly popular today, components of these economic principles have been in use since the 18th and 19th centuries, when the industrial revolution was at its zenith. In the early nineteenth century, manufacturers worked to reduce their negative environmental externalities by modifying industrial practices to create fewer soot or rubbish by-products. In any case, green innovation did not develop as a distinct corporate category until the 1990s. According to a 2018 United Nations report, cumulative global investment in renewable energy and green technology processes surpassed $200 billion in 2017. Since 2004, a total of $2.9 trillion has been invested in renewable energy sources such as solar and wind power. According to the UN, China was the world's top investor in the sector in 2017, with over $126 billion committed (Qamar and Mariya 2021)

"Green technology" is a term that encompasses any technology that has been developed to be environmentally friendly, from its manufacturing process to its applications. Green technology's primary purpose is to conserve the natural environment, regulate climate change, and reduce reliance on non-renewable resources such as fossil fuels, which harm the ecosystem. Energy, waste management, and transportation are among the industries investing in this technology. This technology is useful but must overcome various barriers to function well. This technology has also become a source of employment. Therefore, we must invest in green solutions for human survival and meet technological standards for decreasing environmental risk and saving resources (Qamar and Mariya, 2021).

5. Green Machines and Technology

Green machines and technology provide solutions that reduce the negative impact of technology. It is time to learn from our ancestor's thousands of years ago and utilise renewable resources to generate technology that will be valuable and harmless to the environment. Introducing new products and technology proves the importance of green technology if we want to build a sustainable future (green machines 2020). Green technology saves the environment from destruction. Historically, technology has been blamed for contributing to various environmental issues such as resource depletion, climate change, and pollution. Various environmental issues have been exacerbated by technology. In various cases, the notion of technological development and innovation just does not mix well with environmental preservation. These ideas have existed at opposite poles (Abigail 2021).

5.1. Alternative Sources of Energy

Renewable energy sources: The other alternative energy sources are renewable. These are also referred to as "clean energy." Biogas, geothermal, wind and low-impact hydroelectricity are examples of these. Renewable indicates that they may be found in nature. It produces no greenhouse gas emissions and is associated with no air pollution emissions. Some air pollutants are produced by geothermal energy and biofuels, despite scientists having discovered that pollution from renewable energy sources such as fossil fuels is significantly lower than pollution from non-renewable energy sources. Renewable energy's ability to harness energy from various sources expands the possibilities for long-term sustainability. It also sells at consistent prices. It also encourages large-scale job creation. Fossil fuel technologies do not necessitate a large workforce, but renewable energy necessitates a larger workforce. Growth in the sector could have a favourable "ripple" effect on the local economy (Abigail 2021).

5.2. Server Technology

Server technology brought resource-related challenges with them. The server's price represents a significant capital commitment for the hardware and infrastructure. One of the issues is that they consume a lot of electricity. Most servers also generate eco-waste at the end of their life cycles. Physical servers are also likely to experience data loss during disaster circumstances under their having a physical body that can be ruined through physical means. The answer to these issues would be to create the data equivalent of a server. It is called "cloud computing." Cloud computing works by transferring storage and processing to a digital place on the internet. The creation of a virtual area dramatically reduces energy requirements and usage. It is because there is no hardware to power up. After all, everything is online. It also results in a reduction in carbon emissions. Because there is no hardware, there are no tangible products to wind up as eco-waste. The third major advantage is that cloud computing relies heavily on resource sharing among multiple users. It explains why IT resources are being used more efficiently (Abigail 2021).

5.3. Smart Power Strips

In our digital age, it is not uncommon to discover many devices linked to power outlets. However, most individuals are unaware that devices that are let plugged in consume energy even when they are switched off. These devices are referred to as "energy vampires". They are so-called because they drain the system's power without adding value. Any gadget with an LCD panel or light consumes electricity even when turned off. These include items such as televisions, cable boxes, and microwaves.

In the same way, gadgets that are kept in standby mode continue to consume energy. Traditionally, the only way to prevent power vampires from draining electricity was to unplug the equipment. However, not everyone has the time to go through the entire house looking for unplugged devices. If we agree, smart power strips are an environmentally responsible solution. Smart power strips function by giving us a choice to turn off all gadgets plugged into them. In addition, switching off the receiver on the power strip cuts power to all devices plugged into it (Abigail 2021).

5.4. Energy Efficient Appliances

Electricity is used in a variety of ways by household appliances. However, the devices consume the majority of the household's energy. Dishwashers, refrigerators, and washing machines are examples of these appliances. The most environmentally responsible solution is switching to all devices' eco-friendly versions. It will help to minimise electricity use, resulting in cost savings and lower carbon emissions. The best way to tell if an item is eco-friendly is if it has an Energy Star certification. Energy Star is a long-term initiative of the United States Department of Energy that charges appliances for their energy consumption. The Department of Energy has established criteria for awarding the energy star rating. Standards also differ depending on the type of product. The household device should be more environmentally friendly than previous identical goods (Abigail 2021).

5.5. Telecommuting Software

It is also known as "remote work," a work arrangement in which employees do not have to travel to the office to work. Telecommuting can foster job autonomy, flexibility, and work-life balance. Working from home has environmental advantages. The first benefit is that people do not have to drive to work, reducing the number of cars on the road. It reduces both overall carbon emissions and dangerous gases or pollutants in the air. As a result of digitalisation, paper use is being reduced or abolished.

Furthermore, by not using paper, they are saving trees. As part of the forest ecology, these trees help to reduce carbon emissions and are more helpful in the forest. Telecommuting also helps to reduce the amount of energy utilised by office buildings. These structures frequently consume much more energy than households.

Offices save money on energy bills by allowing employees to work from home. Other options exist for switching to energy-efficient equipment at home (Abigail 2021).

5.6. Device Recycling

Recycling is the re-use of old materials to keep them from becoming waste. It is most likely one of the most important actions everyone can participate in to help the environment. For example, with electronic gadgets, the need to recycle is becoming pressing due to the continuous accumulation of e-waste on a global scale. The advantages of recycling devices include the fact that they first reprocess trash from electronics, which helps to conserve landfill space. Electronics include materials that can be re-used without further processing. It safeguards the materials and energy required for the actual production process. According to Apple, a MacBook Air with a Retina display is 40 per cent recyclable. People can return devices to the company's recycling centre. People who turn in old devices can be rewarded with gift cards—helping the environment by lowering the demand for costly raw materials. Toxic compounds in electronics include mercury, lead, and cadmium, which can leach into water or soil and cause permanent damage. Some states have passed legislation requiring the recycling of outdated devices. However, the more cautiously they are recycled, the more devices join the piles of e-waste that can leak dangerous substances (Abigail 2021).

5.7. Geothermal Heat Pump

People who own heating, ventilation, and air conditioning (HVAC) systems are probably familiar with how they work. It essentially exchanges the air within the house for the air outside. However, it is believed there is an issue with how HVACs operate since they take out warm air in the summer when the air is already warm and suck in warm air in the winter when the temperature outside is below freezing. These air dynamics can cause HVACs to draw more power than necessary. Geothermal heat pumps are an efficient and environmentally beneficial alternative (Abigail 2021).

Geothermal energy is a cost-effective, long-term, and environmentally beneficial heating and cooling technology. Three-fourths of the energy produced by geothermal pumps is derived from the ground. It implies it emits no carbon dioxide or electricity. High-efficiency geothermal systems are roughly 50% more efficient than alternative heating systems such as gas and oil furnaces. A geothermal heat pump functions similarly to an HVAC system. The primary source of electricity for the former comes from underground via an intricate network of underground pipes. During the winter, the temperature outside is often lower than below ground. During this time, the exchangers collect heat from the underground and transport it to the dwelling. During the summer, the fluid in the loop collects heat from the house and returns it underground. During the winter, heat pumps are beneficial for generating hot water within the home (Abigail 2021).

5.8. Closed-Loop Innovation

The whole technology is regulated 'smart systems' monitor uses of resources and management of waste while accounting for metrics which else would never have been comprised (Iravani, Akbari, and Zohoori 2017).

5.9. Green-Chemistry

It is often referred to as "sustainable chemistry." It is a chemical engineering and research philosophy that improves the design of technologies, processes, and products that reduce the use and production of hazardous compounds. Biochemistry, inorganic chemistry, organic chemistry, and analytical chemistry are all examples of green chemistry.

6. Thoughts of Green Technology

"Green" is a label applied to technology and items that are not, in fact, green. So, what exactly is "green technology"? It refers to any technology that seeks to reduce human environmental effects. It includes technology that reduces resource consumption while still incorporating renewable resources. It also includes technology that is non-harmful to the environment. However, there are only a few products in this category (Magazine 2004)

Global warming is a fact and a developing problem that is unsettling society and governments in general, as well as endangering the environment and human health. Green technology is one of the most effective strategies to combat global warming ("10 Examples of Green Technology" 2019)

Scientists and engineers are developing technological solutions to decrease and avoid everything that damages the environment or causes global warming ("10 Examples of Green Technology" 2019)

Numerous instances of green technology are accessible to the average household. The structure and shape of these technologies differ. However, they all share the same goal of reintroducing life into the environment. There is an increasing awareness about environmental matters like global warming. This increased awareness will also enhance the market for environmentally friendly technologies. While some of these items, such as a geothermal pump system or an electric car, can be expensive, their benefits are great. Not everyone will have enough money to choose these possibilities, but there are always more options, to begin with (Abigail 2021).

The aim of green technologies (Iravani, Akbari, and Zohoori 2017) are:

First, to address societal needs in a way that does not deplete natural resources. Second, to create things that can be recycled. Third, to reduce pollution and waste. Finally, to meet the requirements of society in a manner without damaging natural resources.

- To develop products which can be recycled.
- To diminish pollution and waste.
- Use a system that employs cutting-edge approaches to create environmentally friendly products.

The notion of green technologies is eco-friendly, enhanced and utilised in such a manner that it conserves natural resources. Some people use green technology as clean technology. The current expectations are that this field will provide innovation and novelty changes in the life of the same magnitude of information technology. Due to the significance of these technologies, most governments started to enhance them. Thus, government mentions various financial incentives which produce electricity from renewable resources. Various countries look at green technologies to enhance economic growth and develop the lives of their citizens. This technology uses natural and renewable resources, which never ends. This technology also uses innovative techniques and renewable resources in energy production (Iravani, Akbari, and Zohoori, 2017).

6.1. Advantages and Disadvantages of Green Technology

Advantages and disadvantages of green technology (Table 1).

It is difficult to implement green technology; for example, purchasing a hybrid vehicle with good gas efficiency helps minimise energy use; however, hybrid automobiles are frequently more expensive than comparable vehicles without hybrid technology. Though it has importance as we are facing environmental and resources issues. "Green technology" is a system that employs novel approaches to develop ecologically beneficial products. It consists primarily of numerous everyday cleaning products, energy sources, inventions, garbage, apparel, and other items. Going green or employing environmentally friendly technologies are just a few of the many options that countries are considering to boost economic growth and enhance the lives of their populations. Green technology uses renewable natural resources that will never run out. Green technology uses new and innovative energy generation techniques. Green nano-technology that uses green engineering and chemistry is one of the latest green technologies. One of the important factors for environmental pollution is waste disposal (Guo et al., 2020).

Table 1. Advantages and Disadvantages of Green Technology

S.No	Advantages of Green Technology	Disadvantages of Green Technology
1.	It requires minimal upkeep.	Lack of understanding of green technology
2.	It is naturally renewable	Expensive implementation costs
3.	It does not emit any harmful pollutants into the atmosphere.	There are no raw materials or chemicals accessible.
4.	. It generates economic benefits in specific locations.	Uncertainty in performance
5.	It mitigates the effects of global warming by lowering CO_2 emissions.	Lack of human resource skills (Guo et al. 2020)
6.	It is environmentally friendly.	It necessitates a significant investment. Installing new insulation or a roof, for example, to prevent heat from escaping from a home would be considered a green home improvement, but the job would be prohibitively expensive.
7.	It has long-term consequences.	

Making efforts to improve energy efficiency or reduce pollutants caused by households, general living habits, and enterprises is what green technology and processes include. This technology and procedure aim to reduce energy pollution and consumption's potentially negative environmental impact. Green buildings and cars use less energy than fuel, so the initial investment is generally recouped through energy savings over time. The problem is that the savings generated by going green are frequently smaller than expected. They do not compensate for the original expenditure quickly enough to make them economically viable. Going green is also a viable business option, but it must be financially practical (Iravani, Akbari, and Zohoori, 2017).

7. Green Nanotechnology

Green nano-technology utilises green engineering, and chemistry is one of the newest green technologies. One of the significant factors for environmental pollution is waste disposal, which green technology needs to answer well. This technology can alter the waste production and pattern in a manner which it does not worsen the earn, and people can go green, also; among the conceivable fields where these creations and growth are thought to come from compromise green energy, organic farming, eco-textiles, green building construction, and manufacturing of related products and materials to assist green business. Thus, the key goal is to use green technology, which has no negative effects on the planet.

Green nano-technology, which uses green engineering and chemistry, is one of the most modern green innovations. Waste disposal is a key source of pollution in the environment, which green technology must properly address. This technology has the potential to transform waste production and trends in a way that does not impact profits while also allowing people to go green. Green energy, organic farming, eco-textiles, green building construction, and the production of linked products and supplies to enable green business are other probable fields where these breakthroughs and growth are expected to arise. As a result, the major goal is to adopt green technology without negative environmental effects. The use of nanomaterials for purposes such as more efficient solar cells, viable fuel cells, and ecologically friendly batteries is being researched. The most advanced energy-related nano-technology activities are storage, conversion, manufacturing modifications through material and process speed reduction, energy conservation, and increasing renewable energy sources. Due to its unique activity against stubborn pollutants, green nano-technology has considerable potential in water treatment (Qamar and Mariya, 2021).

8. Challenges to Adoption of Green Technology

Because green technology does not account for the environmental costs inherent in many existing production processes, green technology is often more expensive than the technology it wants to replace. This technology is new, and there are several unknowns. Furthermore, the associated training costs and advancements make it even more expensive than other proven technologies. Other considerations, such as technical readiness, supporting infrastructure, human resource competence, and regional features, influence anticipated earnings regarding this technology. Adoption and circulation of these technologies can be rare due to several other hurdles. Some may be institutional, for example, the lack of an appropriate regulatory framework; others can be technological, political, legal, cultural, or financial (Iravani, Akbari, and Zohoori 2017).

Conclusion

Green technology helps to reduce greenhouse gas emissions to zero while still being safe and environmentally friendly. It conserves natural resources and slows environmental degradation. Modern technology has been implemented to protect the environment and conserve resources. It consists of energy-generation procedures, methods, and techniques. Sustainable development is essential to meet the demands of both the current generation and future generations. Carbon dioxide emissions are higher in developed countries compared to developing countries due to extensive industrialisation in countries such as the United States, Russia, and Korea.

Green engineering is founded on the concepts of sustainability, including developing eco-friendly, efficient power use, expanding the life-cycle of products, supporting instruments that indirectly reduce consumption, and restricting the use of hazardous chemicals. Green technology includes energy-saving LED lighting, wind energy, solar panels, composing electric vehicles, programmable thermostats, vertical farming, and the removal of industrial pollution.

The green technology policy aims to provide direction and inspiration so that people can continue to enjoy good quality and a healthy environment. Four pillars should support it. The first goal is to achieve energy independence while reducing the impact on the environment, economy, and social quality of life. Identifying the sectors covered by green technology can be difficult, but our planet is confronting pollution and climate change issues. Because we are aware of the environmental harm, it is self-evident that we must use green technology. One example is the growing desire for electric vehicles (Qamar and Mariya, 2021).

Solar photovoltaic, wind energy, biofuels, biogas, solar thermal, enhanced water mills, and geothermal are all viable green technologies. Therefore, it is suggested that new money streams be generated to fund green technology, new international collaboration on green technologies and an intensification of dialogue on existing national green policies. Furthermore, developing innovative methods and practices that improve green technology is critical. Hence, expand global green energy usage (Guo et al. 2020). Green technology is the need of the future; thus, it is necessary to remove barriers and go ahead.

Chapter 10

Future of Energy Systems

Abstract

Due to scarce energy resources and environmental changes, there will be many future energy system changes. Therefore, it is essential to consider the present for future energy needs. Green energy brought a revolution in the field of the energy system. However, various areas need to be understood to implement green energy successfully. There are many barriers like lack of information, no alternative process of technology, uncertainty about performance, lack of human resource skills, and high implementation cost. Therefore, all countries must make efforts in all sectors like agriculture, infrastructure, food industry, and automotive to overcome these issues. Energy is a world problem; thus, all countries must make integrated efforts in this direction.

This chapter investigates near-future energy technologies and their integration into current infrastructure and their environmental, social, and economic implications to discover solutions to the energy system's current difficulties.

Keywords: energy system, green energy, energy integration, innovation, sustainable energy

1. Introduction

An energy revolution has been initiated today—the major shift from conventional to renewable energy sources. As a result, there is a need for a future energy system that develops technologies emphasising the social, economic, and environmental impacts. In addition, innovative regulations and business models are required to create value (PricewaterhouseCoopers 2021). It also examines their integration into the current infrastructure.

Renewables will account for more than 90% of energy. Furthermore, fossil fuels will account for less than 10% by 2050. Biomass and waste will beat fossil fuels as discoverers of molecules that will be used to heat buildings and power transport and industry. In addition, the capturing and recycling of CO2 will develop novel flows of carbon molecules ("Home" 2022). Future net-zero energy systems will necessitate different ways of doing things, including new technology, consumer habits, collaborations, and regulatory and governance frameworks. The emphasis of work is on understanding how these systems will develop to support partners to understand their role and the benefits of these changes ("Future Energy Systems" 2020). The advantage of using green energy sources is that they must be green or clean; thus, there is no environmental damage. It is also recycled. However, it is not easy to develop green facilities. It also creates economic Benefits for some areas and even progresses the tourism industry (Iravani, Akbari, and Zohoori 2017).

Developing these facilities requires plenty of land, so we might have to cut on agricultural land. Secondly, many green energy sources cannot be installed in particular places on the earth. For example, wave energy can only be prepared if the sea's waves achieve at least 16 feet. The usage of geothermal energy (Figure 16) is only performed in geologically uneven areas of the planet (Figure 17).

Figure 16. Geothermal Earth Crust.
Source: https://cdn.pixabay.com/photo/2014/04/11/16/40/geothermal-energy-321797_960_720.jpg

Areas that cannot use one green energy source method could be substituted for another. For example, if wind generators require more space, they could be set up in the nearby coastline area instead of on land. A study also revealed that electric power could be generated more if it happens in the sea. Research focuses on using various alternative means to generate electric power that requires. An example of this is called ocean thermal energy. This energy is developed by connecting the different temperatures within the water. It is now being serviced on a small scale in Hawaii and Japan. Green energy sources include biodiesel, biomass, geothermal, water, solar, and wind. In the business world, green energy sources can be an attractive goal to achieve goodwill and consumer support. Green improvements should be economically feasible to be implemented successfully (Iravani, Akbari, and Zohoori, 2017).

Figure 17. Geothermal Power Plant.
Source:https://cdn.pixabay.com/photo/2012/11/28/09/18/power-plant-67538__340.jpg

Generally, green technology is more expensive than it aims to use because it accounts for the environmental costs externalised in many conventional production processes. Because it is relatively new, the associated development and training costs can be even more costly than established technologies. The perceived benefits depend on other factors such as supporting infrastructure, technology readiness, human resources capabilities, and geographic elements. Hence, what could be a feasible green technology in one country or region may not be in another. Several other barriers can constrain the adoption and circulation of these technologies. Some may be institutional, such as the lack of an appropriate regulatory framework; others may be technological, financial, political, cultural, or legal. From a company's perspective, the following are likely barriers to adopting green technologies:

- High implementing costs

- Lack of information
- No known alternative chemical or raw material inputs
- No known alternative process technology
- Uncertainty about performance impacts
- Lack of human resources and skills.

2. Steps towards Future Energy System

The 2015 accord and the current net-zero declarations by China and the United States have changed governments from decarbonisation performers to development directors. Massive stimulus packages aimed at spurring a recovery from the pandemic are being implemented from South Korea to Europe and the United States to establish a more sustainable energy system. A new energy system will begin to emerge over the next decade. After being explained, industries such as oil, gas, chemicals, and electricity utilities will merge to develop integrated energy systems ("Home" 2022).

The evolution of tomorrow's energy system is already well underway. In the future, there will be a need for more energy. The work must start today if the energy system is directed at the climate change agenda.

By 2040, demand for the chemicals and electrons that power our world will have increased by 23%. In the foreseeable future, fossil fuels will continue to account for a significant portion of the molecules that drive transportation and the electrons that generate heat and power. In the future, carbon-free energy sources will primarily be electrons, although a growing fraction of molecules will account for a larger share of production.

More than 190 countries have devoted themselves to the goals of the 2015 Paris Agreement, which include aggressive declines in emissions. Investors transferred a record US $350bn into sustainable funds, surpassing the amount in 2019. A large amount of capital is being invested into innovation, research, and development scaling efforts that provide a means to bring down the cost of renewable energy production, energy storage, green hydrogen, and other low-carbon innovations. As coal declines and natural gas generates only modestly after 2030, renewable energy gains market share in the global electricity market. By 2040, growth in solar and wind will account for 47% of the electricity market, up from 29% today.

3. Strategic Implications

It is vital to think strategically to bring sustainability through green energy. There are various strategic implications which are:

3.1. An Increased Need for State Orchestration

There are a variety of obstacles to overcome in many dimensions. The major area to focus on how to costs must be reduced, planned investment infrastructure and evolving risks must be controlled with constantly updated regulatory and operational measures and standards. These attempts necessitate reconsidering the ways and roles of molecules and electrons. They require a reconsideration of the state-market relationship. The transition to net zero and the new energy systems will most likely be managed solely by the government and commercial sector. Consequently, robust state orchestration and joint efforts by the state and market are required to collaborate in novel ways.

Three possible models are predicted to be adopted worldwide in ascending order of degree of state participation: policy driver, co-investor, and strategic infrastructure investor. Because of the complexity and pace of the revolution, all actors, including chemical businesses, national oil companies, service firms, grid operators, and government organisations such as regulators and ministries, must respond strategically. The new path forged by molecules and electrons will achieve more convergence and dismantle conventional boundaries between energy sectors. As a result, companies will be pushed to become more integrated as a result of this. Utilities will need new commercial operations such as data services, analytics, financing and installation of domestic solar panels, charging explanations, and B2B behind the meter energy management ("Home" 2022)

Integrated oil companies and transmission system operators, including OMV, Shell, and TenneT, have invested substantially in the power sector. In addition, chemical companies identify hydrogen as a vital production feedstock and energy source for their applications and business opportunities. As a result, they must reinvent themselves by focusing on harvesting, creating, recycling electrons, and circularly capturing molecules.

3.2. Sustainable Energy Transition

Presently emphasises the following four main issues to permit the transition:

3.2.1. Benchmarking and Country Dialogues
Fast-track transition of national energy systems and significant industrial sectors through fact-based frameworks, tools to benchmark progress of national energy systems, dialogues, and devices to benchmark progress of national energy systems and industrial modification towards net zero emissions, encouraging a broad understanding of priorities and challenges ("Shaping the Future of Energy and Materials" 2019).

3.2.2. Growth Market
Efforts and investments are required to achieve a clean energy transition in countries and regions. It should be done where the economy at large, emissions, and energy demand is rising the most ("Shaping the Future of Energy and Materials" 2019).

3.2.3. Net-Zero Carbon Cities
Bring efforts from the public and private sectors together to assist national and city leaders meet their emission reduction targets by producing an efficient, circular urban energy landscape and electricity ("Shaping the Future of Energy and Materials" 2019).

3.2.4. Transforming Industrial Ecosystem
Combining the strategies and multi-stakeholder coalitions that will provide net-zero material industrial eco-systems with attentive human capital management will enable the design and plan of new approaches towards their future performance ("Shaping the Future of Energy and Materials" 2019).

Permitting businesses, society, and governments to grow sustainable economies helps to stop climate change and create a more equitable world. Evolution of energy, infrastructure, and materials is required in the next two decades to retain global temperature enhancement below 1.5 degrees while safeguarding an energy future that is secure, inclusive, and reasonable. Technologies and new trends are changing energy generation, delivery, and consumption. Correspondingly, world demand for raw materials could double by 2060. Moreover, getting on a path to net-zero emissions by 2050 will require annual investment in clean energy infrastructure to reach nearly $4 trillion by 2030. These steps need a radical restructuring of the economy ("Shaping the Future of Energy and Materials" 2019).

3.3. Future Challenges

3.3.1. Digitalization

The world's energy system will change more in the next ten years than in the previous hundred. The world's energy system is moving from fossil fuels to renewable energy sources. Industrial companies are confronted with addressing this transition through transformative approaches ("Digital in the Future of Energy | GE Digital" n.d.). Digitalisation will make power-producing assets more efficient, the aviation industry more sustainable, the electric grid more secure and resilient, and help manufacturers decrease waste.

3.3.2. Mission of Decarbonization

The next few decades will likely be among the most transformative energy sectors have ever seen. The energy industry is investing billions to confirm that the system is fit for the future. Arup envisages a world with a much more comprehensive and different range of heating sources with lower emissions. Furthermore, new vehicles will be electric. From 2018 to 2035, the potential models will make it possible for the energy market to change to attain a sustainable energy system ("Energy Systems a View from 2035" n.d.).

There will be many interactions and transformation stages from generation to production to implementation. The larger occurrence of recurrent power sources such as solar and wind will produce challenges in aligning peak demand with generation and confirming reliability. It comprises rising demand for green energy and requires the storage of electrons and the requirement of transporting green molecules. New value chains will link distributed power generation to demand centres, storage, recycling, and carbon capture. Implementing new methods to source carbon molecules for industries is another challenge. Solving these challenges increases the conversion and storage of energy, which is essential and will change the conventional role of the carbon molecule.

A new energy system will develop a new method for global energy. Developing renewable electrons and molecules is faking novel trade routes that will require substantial new investment. Projects have been suggested to build electricity transmission networks from areas with an enormous capacity for renewable electricity production to centres of demand. Flows may alter speedily, and participants will fight to achieve the novel energy system equilibrium and the high risks and volatility it will transmit. A power system less dependent on coal and oil will likely present novel geopolitical dynamics. There will be an increasingly global political push for strategic autonomy and self-sufficiency ("Home" 2022).

Carbon Trust assists innovators in developing new value propositions that have disturbed energy markets. For over 18 years, the Carbon Trust has been a trusted partner to an extensive range of organisations grappling with energy matters. For example, the UK power distribution network companies understand how to best integrate electricity storage assets into networks, develop roadmaps for renewable energy technologies and smart architectures for governments worldwide, and measure uncertainty in future energy system business models for multinational corporations.

The need for diverse approaches to deliver net zero energy systems will challenge regulators, governments, consumers, and businesses and their track record of assisting partners in comprehending those challenges ("Future Energy Systems" 2020).

There is a question about why fuels will continue to be a significant element of future energy systems. The reason is the enormous benefits of fuels from their transportation, low cost of storage and handling, and managing the seasonal swing in heating demand for a country having a winter and summer season—for example, the United Kingdom.

The logic of the continuous need for fuels is examined, and synthetic fuels' advantages and disadvantages are weighed against those of fossil fuels. Almost all current energy systems rely heavily on fossil fuels. Part of this is due to the lower cost and accessibility of fossil fuels compared to other significant and secondary energy sources. Still, an often-overlooked factor is an inherent advantage that these fuels provide. The ability to efficiently store terawatt hours (TWh) of chemical energy allows the supply of primary significance to be decoupled from the demand for energy on a large scale in terms of time and location. For a country like the United Kingdom, which has both winter and summer seasons, the ability to store TWh of fuel for heating helps to balance out this significant seasonal demand swing. The shift toward a greater proportion of primary energy derived from weather-dependent renewables (primary electricity) presents challenges in maintaining the balance of demand and supply over a wide range of timescales and distances, with the interseasonal swing in what demand is one of the most significant. The global deployment of weather-dependent renewable production such as solar PV and wind generation appears to continue as costs fall and knowledge of how to successfully integrate the most significant amount of primary electricity within electrical systems (Wilson and Styring 2017).

Renewable output (excluding large hydro) accounted for 45.3, 51.3, and 55.3 per cent of the annual global electrical generation capacity change in 2014, 2015, and 2016, respectively (Krämer n.d.), up from 19.5 per cent in 2007. This level of growth over a decade justifies that weather-dependent renewables can be considered a mainstream technology choice for various countries, which has aided in lowering costs. However, as with other forms of production, there are still numerous areas where technological advancement could bring additional cost enhancement (Wilson and Styring 2017).

The critical historical motive during a time when the costs of solar PV and wind power production were higher was for countries to stimulate investment to reduce carbon emissions. This market expansion was driven by technology and manufacturing advancements that reduced costs, allowing for more investment and growth. As a result, countries must consider their renewable energy resources on a larger scale than previously imagined. As a result, countries signing up to the COP21 Paris Climate Change agreement intend to lower their carbon targets.

For example, in Great Britain, a combination of demand decreases increased low-carbon renewable production. Fuel switching from coal to natural gas production has provided an emissions reduction of 61% in 2016 from a 1990 baseline. With subsidies for renewable production reduction and capacity auctions rising, the direct government cost of subsidising renewable energy is becoming more understood and controlled by the indirect cost of helping higher and higher levels of primary electricity, which still matters to substantial uncertainty.

The increased cost of balancing electrical production and demand in future systems with lower performance for fuel-based production accounts for a significant portion of this. The capability to balance a future energy system is difficult to imagine without the assistance of fuels of some sort. If fossil fuels are inadequate for either climate or other reasons, then there would appear to be strong circumstances for synthetic fuels to take their place. However, it is tricky to conceive of energy systems moving to just-in-time provisions of energy from non-fuel-based primary energy sources to finalise energy demand as the challenge of balancing this over seasons is intractable (Wilson and Styring 2017). This article aims to suggest the continuous, critical requirement for fuel-based storage to overwhelm seasonal changes in energy demands for Great Britain, shown by using multiyear empirical time series data for natural gas, liquid fuels, and electricity. This research proposed that the sheer scale of fuel usage provides the inter-seasonal stores of energy that modern energy systems need. Furthermore, focusing on the use of fuels will still be justified in future energy systems.

The average amount of energy held in storage in the United Kingdom for the main solid (coal), liquid (crude oil, oil products), and gaseous (natural gas) commodities from 2012 to 2016 shows that there is a seasonal variation in the quantity of natural gas and coal held in storage. Still, total inventories of oil products and crude oil have a less seasonal component. It is assumed to be owing to the underlying demand for the fuels themselves, with power (natural gas and coal) and heating (natural gas) having large seasonal fluctuations. In contrast, transportation has fewer (oil products and crude oil). Furthermore, crude oil and oil products are traded globally, importing and exporting considerably more than coal and natural gas.

As Great Britain transfers to securing more and more of its primary energy requirements from renewable electricity sources, there is a shift to weather dependency not only for demand but also for the supply of primary energy. Thus it is said to carry new challenges to balance the differences between demand and supply over diverse timescales, which are undoubtedly factual. However, though energy systems have always needed balancing over various timescales, this is not a new challenge in and of itself. However, the tools and technologies to do so are likely to be. Historically, fuels have permitted this decoupling of primary energy supply from demand on a grand scale at both temporal and scales. The question of whether this average storage stage of energy at 240 TWh is optimised is explored in the next section.

All energy systems benefit from having stores of energy that perform as buffers along their supply chain, from collecting primary energy to the final energy usage. It is driven by a desire for a certain level of energy security, which is a challenging term to theorise. The question is how much energy stored should be accessible to an energy system such as Great Britain's. It is an open research question as it is not optimised at a whole system stage between the natural gas, electrical and liquid fuel networks. Even if the stage of the stored energy of various fuels can be altered comparatively speedily by the difference in demand created by switching fuels. For example, the dash for the gas building of natural gas production in Great Britain (Winskel 2002) had a key impact on the demand for natural gas and coal. According to Wilson and Staffell (cited in Wilson and Styring 2017), the price discrepancy between natural gas and coal as a fuel for electricity production and an effective carbon price offered the market conditions for a fuel switch from coal to natural gas at an unprecedented rate.

There is a massive difference between the two fuels, with natural gas providing 80% of Great Britain's heating demand and primary energy for the electrical system. Coal, in contrast, provides a very slight heating supply and is used in the industrial and electrical sectors. Moving away from natural gas thus carries the grand challenge of decarbonising the heating sector and part of the electrical sector. However, assuming that the stored energy of coal is inaccessible in Great Britain's future energy system (without CCS). The same may be contented for natural gas, too, as this will come under increasing pressure to move off the system into an intensively decarbonised future.

If natural gas and coal are no longer accessible to Great Britain, it is open research query how it would balance its energy demand over the year. Thus, due to resource constraints, it is problematic to imagine tens or hundreds of TWhs of energy can be stored in any energy form that is not a fuel. Non-fuel approaches to storing energy are orders of magnitude more costly than storing fuels. Though they will have significant roles to play in terms of harmonising shorter and medium-term demand and supply (within day and days to weeks), they are ill-matched to provide TWh levels of seasonal harmonising needed by heating demand in specific.

Assuming that coal's stored energy is thus unreachable in Great Britain's future energy system (in the absence of CCS), the same may be true for natural gas, which will face increasing pressure to exit the system aggressively decarbonised future. There is a significant disparity between the two fuels, with natural gas providing 80 per cent of Great Britain's heating demand and primary energy for the electrical system. On the other hand, coal provides only a minor heating supply and is mostly employed in the industrial and electrical sectors. Moving away from natural gas poses the major task of decarbonising the heating and power sectors.

Conclusion

This chapter investigates near-future energy technologies and their integration into current infrastructure and their environmental, social, and economic implications to discover solutions to the energy system's current difficulties. An energy revolution has been initiated today—the major shift from conventional to renewable energy sources. Innovation and developing regulations and business models are required to create value. The strategic implications for the future energy system are an increased need for state orchestration and the mission of Decarbonization and sustainable energy transition. As a result, the world's energy system will change more in the next ten years than in the previous hundred.

There will be many interactions and transformation stages from generation to production to implementation. The larger occurrence of recurrent power sources such as solar and wind will produce challenges in aligning peak demand with generation and confirming reliability. It comprises rising demand for green energy and requires the storage of electrons and the requirement to transport green molecules. New value chains will link distributed power generation to demand centres, storage, recycling, and carbon capture.

The logic of the continuous need for fuels is examined. Furthermore, the advantages and disadvantages of synthetic fuels are weighed against those of fossil fuels. Almost all current energy systems rely heavily on fossil fuels. Part of this is due to the lower cost and accessibility of fossil fuels compared to other major and secondary energy sources. Still, an often-overlooked factor is an inherent advantage that these fuels provide. The focus of the world is to make it pollution free. The ultimate solution is to adopt a clean energy system. The energy system should develop innovative solutions and also consider current problems. Various countries are making efforts in this direction. However, there is no proper alignment between policies and the implementation process. There is a need for collaborative efforts of people from all countries, sectors, and organisations to cope with the situations.

References

Abigail. 2021. "*11 Green Technology Applications in Everyday Life.*" Robots.net. May 10, 2021. https://robots.net/tech/green-technology-applications-in-everyday life/#:~:text= Most%20of%20the%20applications%20of.

ADB head honcho. 2010. "*Developing Smart Grid Technology for Efficient Utilization of Renewable Energy.*" Asian Development Bank. December 8, 2010. https://www.adb.org/projects/43053-012/main.

Akadiri, Peter O., Ezekiel A. Chinyio, and Paul O. Olomolaiye. "Design of a Sustainable Building: A Conceptual Framework for Implementing Sustainability in the Building Sector." *Buildings* 2, no. 2 (May 4, 2012): 126–52. https://doi.org/10.3390/buildings 2020126.

Akadiri, Peter O., Ezekiel A. Chinyio, and Paul O. Olomolaiye. 2012a. "Design of aSustainable Building: A Conceptual Framework for Implementing Sustainability in the Building Sector." *Buildings* 2 (2): 126–52. https://doi.org/10.3390/buildings 2020126.

ARIES: Advanced Research on Integrated Energy Systems. n.d. Www.nrel.gov. https://www.nrel.gov/aries/.

Bressand, Florian, Dianna Farrell, and Pedro Hass. 2007. "*McKinsey Global Institute Curbing Global Energy Demand Growth: The Energy Productivity Opportunity.*" https://www.mckinsey.com/~/media/mckinsey/business%20functions/sustainability/our%20insights/curbing%20global%20energy%20demand%20growth/mgi_curbing_global_energy_demand_full_report.pdf.

Burton, Larry. "*Examples of Sustainable Development in the U.S.*" resource.temarry.com, July 25, 2021. https://resource.temarry.com/blog/examples-of-sustainable-development-in-the-us.

Cambridge Dictionary. 2020. "*ENERGY | Meaning in the Cambridge English Dictionary.*" Cambridge.org. January. https://dictionary.cambridge.org/dictionary/english/energy.

Clean Energy Integration. n.d. INL. https://inl.gov/research-program/clean-energy-integration/

Davies, Laura. "*Types and Alternative Sources of Renewable Energy.*" EDF Energy. EDF Energy, December 21, 2017. https://www.edfenergy.com/for-home/energywise/renewable-energy-sources.

Davies, Laura. 2017. *Types and Alternative Sources of Renewable Energy*. EDF Energy. EDF Energy. December 21, 2017. https://www.edfenergy.com/for-home/energywise/renewable-energy-sources.

Digital in the Future of Energy | GE Digital." n.d. www.ge.com. Accessed, 2022. https://www.ge.com/digital/future-of-energy?utm_medium=Paid-Search&utm_source=Google&utm_campaign=Brand-EnergyTransition-TOF-US-Search&utm_content=future%20energy%20system.

Digital in the Future of Energy | GE Digital." n.d. Www.ge.com. https://www.ge.com/digital/future-of-energy?utm_medium=Paid-

Search&utm_source=Google&utm_campaign=Brand-EnergyTransition-TOF-US-Search&utm_content=energy%20renewables.

Dorsey, Piccirilli. 2019. *"Energy Efficiency | EESI."* Eesi.org. 2019. https://www.eesi.org/topics/energy-efficiency/description.

Efficient Energy Use. 2022. *Wikipedia.* April 12, 2022. https://en.wikipedia.org/wiki/Efficient_energy_use#:~:text=Efficient%20energy%20use%2C%20sometimes%20simply.

Energy - Definition, Meaning & Synonyms. n.d. Vocabulary.com. Accessed May 6, 2022. https://www.vocabulary.com/dictionary/energy.

Energy Efficiency and Conservation - U.S. Energy Information Administration (EIA). 2020. Www.eia.gov. December 8, 2020. https://www.eia.gov/energyexplained/use-of-energy/efficiency-and-conservation.php.

Energy Systems a View from 2035. n.d. Www.arup.com. Accessed, 2022. https://www.arup.com/perspectives/publications/research/section/the-future-of-energy-2035.

Energy Utilization Promotion of Energy Saving (Policy). n.d. Accessed May 1, 2022. https://www.scj.go.jp/ja/info/kohyo/pdf/kohyo-20-t34-1s5e.pdf.

Five Reasons Integrated Energy Is so Important. 2018. Sunny. SMA Corporate Blog. July 17, 2018. https://www.sma-sunny.com/en/five-reasons-integrated-energy-is-so-important/.

Future Energy Systems. 2020. Https://Www.carbontrust.com/What-We-Do/Future-Energy-Systems. January 28, 2020. https://www.carbontrust.com/what-we-do/future-energy-systems.

Global Green Growth Institute-*A Resilient World Through Inclusive and Sustainable Green Growth*, https://gggi.org/site/assets/uploads/2017/12/GGGI%E2%80%99s-Technical-Guidelines-on-Green-Energy-Development_dereje-senshaw2017.pdf.

Green Energy Systems, *Indore - Manufacturer of Solar Product and Solar Security System.* n.d. IndiaMART.com. Accessed May 4, 2022. https://www.indiamart.com/green-energysystems/.

Green Energy Systems." 2015. Green-Energysystem.com. 2015. https://www.green-energysystem.com/.

Greenmachines. 2020. *"The Importance of Green Technology."* Green Machines. August 4, 2020. https://greenmachines.com/the-importance-of-green-technology/

Guo, Minjian, Joanna Nowakowska-Grunt, Vladimir Gorbanyov, and Maria Egorova. 2020. "Green Technology and Sustainable Development: Assessment and Green Growth Frameworks." *Sustainability* 12 (16): 6571. https://doi.org/10.3390/su12166571.

Gwen, Farnsworth. 2015. *"Renewable Energy and Energy Efficiency."* Western Resource Advocates. 2015. https://westernresourceadvocates.org/clean-energy/renewable-energy/.

Hole, Tom. 2017. *"What Are the Most Efficient Forms of Renewable Energy."* Born to Engineer. August 2, 2017. https://www.borntoengineer.com/efficient-form-renewable-energy.

Home and Business Energy Systems | *Green Energy Systems."* 2020. December 30, 2020. https://greenenergysystems.uk/.

References

Home. 2022. www.futureenergysystems.ca. February 3, 2022. https://www.futureenergy systems.ca/#:~:text=Future%20Energy%20Systems%20develops%20energy.

Home. n.d. Integrated Energy Network. http://integratedenergynetwork.com/

Integrated Energy Systems Home. n.d. Ies.inl.gov. https://ies.inl.gov/SitePages/Home.aspx#:~:text=A%20power%20plant%20being%20used.

Integrated Energy Systems. n.d. Aalborgcsp.com. https://www.aalborgcsp.com/business-areas/integrated-energy-systems/.

Integrated Energy Systems. n.d. Www.danfoss.com. https://www.danfoss.com/en/about-danfoss/insights-for-tomorrow/integrated-energy-systems/.

Iravani, Abolfazl, Mohammad Hasan akbari, and Mahmood Zohoori. 2017. "Advantages and Disadvantages of Green Technology; Goals, Challenges and Strengths." *International Journal of Science and Engineering Applications* 6 (9): 272–84. https://doi.org/10.7753/ijsea0609.1005.

Irsyad, M. Indra al, Anthony Basco Halog, Rabindra Nepal, and Deddy P. Koesrindartoto. 2017. "Selecting Tools for Renewable Energy Analysis in Developing Countries: An Expanded Review." *Frontiers in Energy Research* 5 (December). https://doi.org/10.3389/fenrg.2017.00034.

Ismail, M.S., M. Moghavvemi, T.M.I. Mahlia, K.M. Muttaqi, and S. Moghavvemi. 2015. "Effective Utilization of Excess Energy in Standalone Hybrid Renewable Energy Systems for Improving Comfort Ability and Reducing Cost of Energy: A Review and Analysis." *Renewable and Sustainable Energy Reviews* 42 (February): 726–34. https://doi.org/10.1016/j.rser.2014.10.051.

Johannsen, Rasmus Magni, Poul Alberg Østergaard, David Maya-Drysdale, and Louise Krog Elmegaard Mouritsen. 2021. "Designing Tools for Energy System Scenario Making in Municipal Energy Planning." *Energies* 14 (5): 1442. https://doi.org/10.3390/en14051442.

Kalyani, Vijay Laxmi, Manisha Kumari Dudi, and Shikha Pareek. "Green Energy: The Need of the World." *Journal of Management Engineering and Information Technology* 2, no. 5, ISSN No 2394-8124, (October 5, 2015): 18–26.

http://www.jmeit.com/JMEIT%20Vol%202%20Issue%205%20Oct%202015/JMEITOCT0205005.pdf.

Krämer, Johannes. n.d. "Home." *Frankfurt School*. Accessed April, 2022. https://www.fs-unep-centre.org/.

Li, Peng, Fan Zhang, Xiyuan Ma, Senjing Yao, Zhuolin Zhong, Ping Yang, Zhuoli Zhao, Chun Sing Lai, and Loi Lei Lai. 2021. "Multi-Time Scale Economic Optimization Dispatch of the Park Integrated Energy System." *Frontiers in Energy Research* 9 (September). https://doi.org/10.3389/fenrg.2021.743619.

Liu, Zhenya. "Global Energy Development: The Reality and Challenges." *Global Energy Interconnection*, 2015, pp. 1–64., https://doi.org/10.1016/b978-0-12-804405-6.00001-4. https://doi.org/10.1016/B978-0-12-804405-6.00001-4Get rights and content.

Magazine, From the Editors of E. 2004. "*Nader in Ruins*." Emagazine.com. October 19, 2004. https://emagazine.com/an-introduction-to-green-technology/n.

Mcfarland, Kevin. 2017. "SynTech Bioenergy." *SynTech Bioenergy*. October 24, 2017. https://www.syntechbioenergy.com/blog/biomass-advantages-disadvantages.

Moe Long. *"What Is Green Technology and Examples of Its Benefits?"* www.electropages.com, September 27, 2019. https://www.electropages.com/blog/2019/09/what-is-green-technology.

Mueller, S. 2017. *"Green Technology and Its Effect on the Modern World."* Undefined. https://www.semanticscholar.org/paper/Green-Technology-and-its-effect-on-the-modern-world-Mueller/22e3ca91bf6263834ee2da52fbc06827f5e33431.

O'Malley, Mark, Benjamin Kroposki, Bryan Hannegan, Henrik Madsen, Mattias Andersson, William D'haeseleer, Mark F. McGranaghan, et al. 2016. *"Energy Systems Integration. Defining and Describing the Value Proposition,"* June. https://doi.org/10.2172/1257674.

O'Neil, Neasan. *"A Low Carbon Future Needs an Integrated Energy System, Say Imperial Researchers | Imperial News | Imperial College London."* Imperial News, 26 Apr. 2018, www.imperial.ac.uk/news/185893/carbon-future-needs-integrated-energy-system/. Accessed April, 2022.

Pathways to Sustainable Energy Accelerating Energy Transition in the UNECE Region.' n.d. https://unece.org/DAM/energy/se/pdfs/CSE/Publications/Final_Report_PathwaysToSE.pdf.

Power Utilization and Energy Efficiency." n.d. accessed, 2022. https://caper-usa.com/research-program/power-utilization-and-energy-efficiency/.

PricewaterhouseCoopers. 2021. *"Energy Transition: Tomorrow's Energy System."* PwC. November 9, 2021. https://www.pwc.com/gx/en/industries/energy-utilities-resources/future-energy/inventing-tomorrows-energy-system.html.

Prindle, Bill, and Maggie Aldridge. *"The Twin Pillars of Sustainable Energy: Synergies between Energy Efficiency and Renewable Energy Technology and Policy."* ACEEE Report Number E074, May 2017.

Qamar, Muhammad Zaid, and Noor Mariya. 2021. "Green Technology and Its Implications Worldwide." *The Inquisitive Meridian* 3 (1).

Salvarli, Mustafa Seckin, and Huseyin Salvarli. 2020. Review of For Sustainable Development: Future Trends in Renewable Energy and Enabling Technologies. In *Renewable Energy - Resources, Challenges and Applications*, edited by Mansour Al Qubeissi. IntechOpen.

Shafiei, M., and H. Abadi. 2017. *"The Importance of Green Technologies and Energy Efficiency for Environmental Protection."* www.semanticscholar.org. 2017. https://www.semanticscholar.org/paper/The-Importance-of-Green-Technologies-and-Energy-for-Shafiei-Abadi/ce752d72a6a82fc5e51f3d719f77377c119582bb.

Shah, Kashish. 2021. *Ieefa.org.* IEEFA. https://ieefa.org.

Silvast, Antti, Simone Abram, and Claire Copeland. 2021. "Energy Systems Integration as Research Practice." *Technology Analysis & Strategic Management,* September, 1–12. https://doi.org/10.1080/09537325.2021.1974376.

Sustainable Design. n.d. Www.gsa.gov. https://www.gsa.gov/real-estate/design-and-construction/design-excellence/sustainability/sustainable-design.

Ten Examples of Green Technology. 2019. TECAM. June 18, 2019. https://tecamgroup.com/10-examples-of-green-technology/

The Social Benefits of Sustainable Design, n.d. https://www1.eere.energy.gov/femp/pdfs/buscase_section3.pdf.

References

Tomar, Pradeep. 2021. "Role of Renewable Energy Techniques to Design and Develop Sustainable Green Building." *Research Anthology on Clean Energy Management and Solutions*, 1185–97. https://doi.org/10.4018/978-1-7998-9152-9.ch051

TWI. n.d. "*What Is Green Energy?* (Definition, Types and Examples)." www.twi-Global.com. https://www.twi-global.com/technical-knowledge/faqs/what-is-green-energy. 99, 681-692.

United Nations. 2018. "*Energy - United Nations Sustainable Development.*" United Nations Sustainable Development. 2018. https://www.un.org/sustainable development/energy/.

Vezzoli, Carlo, Fabrizio Ceschin, Lilac Osanjo, Mugendi K. M'Rithaa, Richie Moalosi, Venny Nakazibwe, and Jan Carel Diehl. Designing Sustainable Energy for All. *Green Energy and Technology*. Cham: Springer International Publishing, 2018. https://doi.org/10.1007/978-3-319-70223-0.

What Is Green Energy? Renewable Energy Source. n.d. Justenergy.com. https://justenergy.com/blog/what-is-green-energy/.

Wikipedia Contributors. 2019. "Green Building." *Wikipedia*. Wikimedia Foundation. April 19, 2019. https://en.wikipedia.org/wiki/Green_building.

Wilson, I. A. Grant, and Peter Styring. 2017. "Why Synthetic Fuels Are Necessary in Future Energy Systems." *Frontiers in Energy Research* 5 (July). https://doi.org/10.3389/fenrg.2017.00019.

World Economic Forum. 2019. *Shaping the Future of Energy and Materials* https://www.weforum.org/platforms/shaping-the-future-of-energy.

World Green Building Council. 2016. "*What Is Green Building?* | World Green Building Council." Worldgbc.org. 2016. https://www.worldgbc.org/what-green-building.

About the Author

Dr Teena Mishra has 11 years of work experience. She worked as a lecturer, assistant professor, and manager in various organizations. She recently worked with PSCIVE in Bhopal, received her PhD from Barkatullah University in Bhopal, and completed PG in human resource management, PGJMC, and B. A in advertising and sales promotion and has participated in various training programs and other academic profiles. In addition, she engaged in various extracurricular activities at the institute. She has experience in various fields like teaching, promotional activities, administration, creative writing, and research. She has taught undergraduate and post-graduate courses and various subjects like human resource management, organizational behaviour, marketing management, and retail management. Her areas of research interest include sustainable development and the environment; sustainable practices; organizational behaviour; internal marketing; the educational system; marketing practices and strategy; and human resource practices.

She has presented various research papers at national and international conferences. In addition, she has had her work published in several international peer-reviewed journals, including the International Journal of Business and Commerce Management, the International Journal of Creative Research Thoughts, the Unnyan Journal, Springer publication, the International Journal of Trends in Scientific Research and Development, and the Kavv International Journal of Economics, Commerce, and Business, among others.

Index

B

biomass, vi, 5, 6, 15, 18, 23, 38, 39, 40, 41, 44, 45, 50, 53, 54, 56, 62, 63, 64, 70, 98, 118, 119, 147, 149, 161

C

clean energy, vi, 2, 3, 5, 8, 13, 17, 30, 35, 37, 40, 41, 44, 55, 58, 60, 63, 64, 66, 68, 69, 91, 102, 103, 105, 108, 122, 136, 152, 158, 159, 163

E

efficient energy system, 107, 118
energy conservation, 14, 38, 46, 71, 84, 98, 107, 112, 113, 122, 143
energy system integration, vi, 17, 19, 31, 32, 35
energy utilization, vi, 75, 160
energy utilization system, vi

F

forms of green energy, 54, 62, 64
future energy system, vii, 32, 147, 148, 150, 154, 155, 157, 158, 160, 163

G

geothermal energy, vi, 39, 41, 54, 56, 58, 59, 62, 63, 64, 104, 118, 119, 136, 139, 148

green building, vi, vii, xi, 51, 65, 72, 73, 74, 77, 83, 84, 95, 100, 101, 102, 104, 116, 125, 142, 143, 163
green energy, ii, v, vi, 1, 2, 5, 6, 7, 8, 10, 12, 13, 14, 15, 37, 38, 39, 40, 41, 42, 44, 47, 48, 49, 50, 51, 53, 54, 55, 56, 57, 60, 62, 63, 64, 66, 75, 77, 78, 79, 80, 81, 82, 91, 92, 93, 122, 123, 143, 145, 147, 148, 151, 153, 158, 160, 161, 163
green energy products, 49, 57
green energy system, 1, 37, 51, 77, 78, 79, 80, 81, 82, 91, 92, 160
green technology, ii, v, vii, 51, 123, 124, 125, 126, 127, 128, 129, 130, 131, 132, 133, 134, 135, 136, 140, 141, 142, 143, 144, 149, 159, 160, 161, 162
green utilization system, vi

H

hydroelectric energy, 53

I

integrated energy network, 17, 34, 36, 161

M

modern appliances, vii, 65, 66, 75, 107, 113, 114, 122

Index

R

renewable energy, vi, xii, 1, 3, 4, 5, 6, 9, 11, 12, 14, 15, 17, 18, 19, 21, 23, 24, 25, 26, 27, 28, 34, 35, 37, 38, 39, 40, 41, 42, 43, 44, 45, 46, 47, 48, 49, 50, 53, 54, 55, 56, 57, 58, 60, 62, 64, 65, 66, 67, 68, 70, 71, 80, 82, 83, 84, 85, 86, 87, 88, 91, 92, 96, 99,100, 103, 107, 110, 114, 118, 119, 120, 121, 125, 127, 133, 135, 136, 143, 147, 150, 153, 154, 155, 157, 159, 160, 161, 162, 163

S

sector integration, vi, 17, 27, 29, 36, 83
solar energy, 10, 15, 18, 39, 40, 41, 44, 53, 54, 55, 58, 60, 62, 63, 78, 85, 91, 100, 103, 119, 128, 129, 133
sustainable development, vi, 9, 13, 15, 37, 38, 42, 43, 44, 45, 46, 47, 51, 66, 68, 69, 74, 85, 92, 93, 96, 99, 105, 125, 144, 159, 160, 162, 163, 165
sustainable energy, vi, vii, 15, 17, 26, 27, 35, 42, 43, 45, 47, 51, 53, 57, 65, 66, 67, 68, 69, 70, 71, 72, 75, 95, 96, 98, 99, 107, 123, 124, 147, 150, 153, 158
system design, v, vi, 74, 77, 79

T

technology advancement, 95
tidal energy, vi, 5, 53, 54, 56

U

urbanization and sustainability, vi

W

wind energy, vii, 10, 18, 28, 39, 41, 53, 54, 55, 60, 63, 104, 118, 128, 144